EXPLORING THE MATH AND ART CONNECTION

EXPLORING THE MATH AND ART CONNECTION

Teaching and Learning Between the Lines

Daniel Jarvis and Irene Naested

Brush Education

Brush Education Inc.
Calgary, Alberta, Canada
www.brusheducation.ca
contact@brusheducation.ca

Library and Archives Canada Cataloguing in Publication
Jarvis, Daniel
Exploring the math and art connection : teaching and learning between the lines / Daniel Jarvis and Irene Naested.
Includes bibliographical references.
Issued also in an electronic format.
ISBN 978-1-55059-398-3
1. Mathematics in art--Textbooks. I. Naested, Irene Mae
II. Title.
N72.M3J37 2012 700.1'08 C2011-908482-1

Printed and manufactured in Canada.

Produced with the assistance of the Government of Alberta, Alberta Multimedia Development Fund. We also acknowledge the financial support of the Government of Canada through the Canada Book Fund for our publishing activities.

Government of Alberta

Canadian Heritage Patrimoine canadien

Acknowledgements

We would like to thank the art education instructors—Irene Naested, Connie Ptasinski, Tonya Seyer, Christa Volk—and their students at Mount Royal University (Calgary, AB), as well as Daniel Jarvis' art education students at Nipissing University (North Bay, ON).

We would also like to express appreciation to the teachers and students at the Calgary Science School (Calgary, AB) and at Timiskaming District Secondary School (New Liskeard, ON) for their creative artwork contributions to this book project.

Finally, we would like to thank Mount Royal University, Nipissing University, and the Canadian Society for Education Through Art (CSEA-SCÉA) for their generous financial contributions that have enabled us to produce this text using colour printing.

Table of Contents

1. Mathematics and Visual Arts Education /9
2. Understanding the Language of Mathematics and Visual Arts /36
3. Mathematics and Art in Flora and Landscapes /55
4. Mathematics and Art in Fauna of the Land, Sea, and Skies /97
5. Mathematics and Art in the Human Figure /123
6. Mathematics and Art in the Designed Environment /164
7. Mathematics and Art in Designed Objects /202
8. Making Connections: Teaching and Learning Between the Lines /235

References /252
Appendix A: Glossary of Key Mathematics Terms /258
Appendix B: Glossary of Key Visual Arts Terms /264
Appendix C: Grade 9 Integrated Project Sample–Golden Ratio & Proportion /271
Appendix D: Grade 12 Integrated Project Sample–Fluxions & Deductions (Calculus) /277
Appendix E: Other Recommended Print and Online Resources /286
Index of Learning Experiences /292
Image Credits /296
About the Authors /304

1 Mathematics and Visual Arts Education

Teaching and Learning Between the Lines

A line defies simple definition. Within mathematics, it has been described as a geometrical object having length but no depth or width formed by a straight set of points that extends to infinity in both directions, and as a set of the points whose coordinates satisfy a given linear equation on the Cartesian plane or in Euclidean space. In visual arts, the line has been referred to as a "basic element," often defined as a continuous mark made on a surface. In art, a line may be straight, curved, bent, thick, thin, broken, vertical, horizontal, or freehand; it is a tool used to visually communicate patterns, two-dimensional shapes, and three-dimensional spaces or objects. The study and use of "line" has also formed a major component, with related skills, of both the mathematics and visual arts curricula in formal schooling.

Teaching and learning "between the lines," metaphorically speaking, involves the recognition of this similarity and the celebration of this and many other connections that exist between these two rich disciplines. There are many

C1-1 Iris images repeated in various media (colour plate i)

reasons why it is beneficial to connect mathematical concepts with visual arts education. This chapter presents how individuals develop both mathematical and visual literacies, and how these two academic journeys can and do intersect and overlap. Herein we discuss teaching and learning theories and strategies that support learning experiences that explore in significant ways these important connections. These types of pedagogical approaches include constructivism, problem-based learning, interdisciplinary learning, brain-based learning, differentiated instruction, and universal design for learning. This chapter also highlights the positive and synergetic effects of integrating mathematics and visual arts on the overall educational experience of children. We briefly present how mathematics and art phobias, or anxieties, are perpetuated and what can be done to offset these negative realities for teachers, parents, and most importantly, students.

Two Significant Perspectives on the World

> Why do many flowers have five or eight petals, but very few six or seven? Why do snowflakes have six-fold symmetry? Why do tigers have stripes but leopards spots? Throughout human history, artists and scientists alike have been inspired by the form and beauty of the natural world. Our changing vision of the universe, and of our place within it, reflects an ever-growing understanding of pattern and structure in nature. Human mind and culture have invented a formal system of reasoning that lets us recognize, classify, and exploit patterns, whatever they may be and wherever they arise. (Stewart, 1995, cover overleaf introduction)

According to Stewart, this formal system of reasoning is called *system mathematics*. Others have described the interpretation of the natural world as art, or aesthetic awareness. Are both types of perspectives possible, valid, and widely available? Doczi (1981) presents a less systematic approach to understanding the visual world, as characterized in the following aesthetically inspired description of similar natural phenomena:

> Why do apple blossoms always have five petals? Only children ask such questions . . . When we look deeply into the patterns of an apple blossom, a seashell, or a swinging pendulum, however, we discover a perfection, an incredible order, that awakens in us a sense of awe that we know as children. (p. i)

Children enjoy exploring natural and human-designed environments through their senses of sight, sound, smell, taste, and touch. These experiences can teach a great deal about how and why things are put together. Children can learn about many mathematical concepts, as well as the elements of visual composition, or design, which comprise line, shape, texture, form, tone (value), and colour.

C1-2 Flowers with five petals (colour plate i)

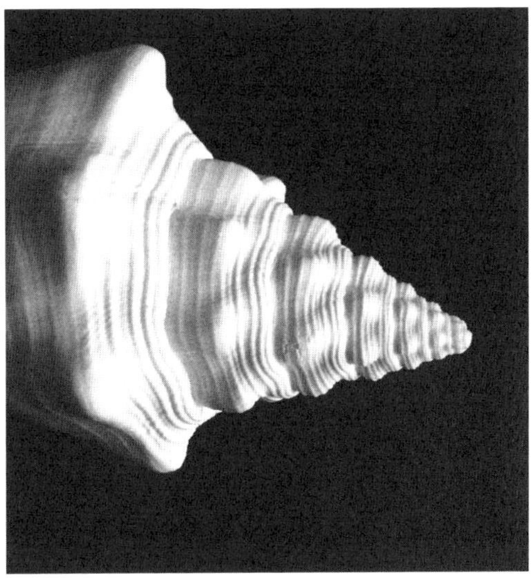

C1-3 Seashell

These elements are to be found in natural forms such as bees' nests, seashells, spider webs, flowers, and bird wings, to name but a few. They are also the elements that have been mathematically organized into human-designed forms. For example, numbers and shapes were represented in the Lascaux paintings of early cave artists in France and Spain. They are, mathematically speaking, two-dimensional maps of objects in space; yet they are also historic and beautiful.

Some Historical Connections

Culture, Mathematics, and Visual Art

Many writers, scientists, and mathematicians give credit to the Egyptian, Greek, and Roman cultures for their strong influence on our present understanding of mathematics, art, and architecture. While there are obviously many contributions made by the above-mentioned cultures, some of which we shall highlight in this chapter, we also acknowledge that there are many Indigenous world cultures that have also demonstrated a strong understanding of mathematics in their writing, artworks, and built structures/environments.

Indigenous peoples have continually looked to nature for guidance and inspiration when creating images, functional forms, and architectural designs; hence a number of these connections will also be made explicit throughout the text within the descriptions of the various learning experiences. Ethnomathematics is a branch of mathematics education dealing with the study of the relationship between mathematics and culture (see, for example, D'Ambrosio (1985); Ascher (1991); Jarvis & Namukasa (2009)). In terms of the relationship between art education and culture, Freedman (1987) provides an expansive overview in her book *Art Education as Social Production: Culture, Society, and Politics in the Formation of Curriculum.*

C1-4 First Nations teepee (colour plate i)

C1-5 West Coast totem sculpture of human figure (colour plate i)

The Ancient Greeks

Mathematics and the visual arts have coexisted since the dawn of human history. Furthermore, beyond mere coexistence, they have been intricately interwoven through issues of form and function throughout every cultural era. In ancient Greece, sculptors such as Phidias, designer of the Parthenon frieze panels and the monumental ivory and gold statue of Athena for which the temple was built, used the *golden ratio* extensively in their work. This mathematical proportion was believed by the Greeks to hold

C1-6 Greek Parthenon (colour plate ii)

the profound secret of visual harmony in the universe. The legacy of this concept, which would later be referred to as the "divine proportion" during the Renaissance, is still evident in both modern and postmodern styles of architecture and fine art. Alex Colville, one of Canada's most widely recognized artists and who has used the golden ratio in his compositions, described the aesthetic experience involved:

> Once you begin to perceive these relationships, circles, spirals, triangles and rectangles appear as if on their own. The beauty of it comes as a surprise, and its harmonies inspire joy. Part of the excitement is that these discoveries, though new to us, are about immutable laws that have been in force since time and space began. (cited in Fry, 1994, p. 35)

Throughout history, artists have made use of many other mathematical elements, including "the geometric shapes, symmetry, the earth's measurements, the proportions of humankind, the patterns of the stars, conic sections [circle, ellipse, parabola, hyperbola], as well as the computer" (Attenborough, Pattison, Patsiatzis, & Muller, 1997). Newman and Boles (1992) maintain that although these two disciplines are often viewed as polarities, they are in fact "the left and right hand of cultural advance: one is the realm of metaphor, the other, the realm of logic . . . Our humanness depends upon a place for the fusion of fact and fancy, emotion and reason. Their union allows the human spirit freedom" (p. xiv).

Socrates, Plato, and Aristotle represent a rich lineage of effective pedagogical transfer. All three of these teachers held to strong, but distinct, educational philosophies. Socrates, master of the strategic question-and-answer method that bears his name, was convinced that education and healing were closely related, and therefore defined teaching as the "building, in a pupil, of a system of value priorities and preferences that defined the healthy soul" (Broudy & Palmer, 1965). Socrates was also one of the first Western educators to hold the belief that unless citizens had an understanding of art and music, they were not considered to be adequately educated (Naisbitt & Aburdene, 1990). In *The Republic*, Plato, using his metaphor of the divided line, describes four "states of mind" or "ways of knowing." The lower levels of illusion (*eikasia*) and belief (*pistis*) encompass the "distorted perceptions" found within poetry and art (trans. 1955/1987, p. 316). To the third level of reason (*dianoia*) belong mathematics and the sciences. "For Plato the study of mathematics was to shape the soul, just as music and literature shape the soul" (Broudy & Palmer, 1965, p. 44). Only at the fourth and highest level, known as intelligence (*noesis*), is the clearest mental vision and recognition of truth achieved through dialectic (pp. 41–42).

Whereas both Socrates and Plato acknowledged the important, yet separate and unequal, emphases on mathematics and education in the arts, it was Aristotle, the most prized student of Plato's Academy, who is remembered for encouraging the cognitive "search for relationships between things apparently disconnected"

(Newman & Boles, 1992). And so, it is this latter quest, no doubt influenced in part by his predecessors and their teachings, that bears directly on an integrated approach to the two disciplines.

For the Greeks of the ancient world, much of their mathematics dealt with geometry (a word derived from "land measure"), a topic that had been explored by and learned from the Egyptians. Euclid of Alexandria, often referred to as the "father of geometry," founded a school in the Hellenistic period and collected many of the then-existing mathematics manuscripts into a carefully organized, thirteen-volume set known as *The Elements*. This popular text would be widely consulted by mathematicians for the next 2,000 years.

Finally, one would be remiss in discussing Grecian influence on interdisciplinary education without mentioning he who has been referred to as the father of both mathematics and music, Pythagoras. It was his passionate quest for learning in many different areas that inspired his disciples to become members of his covert mathematical brotherhood and to continue in his pursuits after his death. Many of the philosophical, mathematical, and aesthetic pursuits of the ancient Greek culture would be revisited and expanded over a thousand years later during the European Renaissance.

The Renaissance

The rebirth of culture and learning that pervaded Europe during the Renaissance was intrinsically tied to the notion of the individual as brave and enlightened adventurer. This social metaphor had, of course, its literal parallels in the likes of explorers Columbus and Magellan, who actually crossed unknown waters in search of the affluent Far East and international glory. But perhaps more fascinating still was the emergence of the artist/scientist who pushed the known boundaries of learning and discovery to new limits. De La Croix and Tansey (1986) highlight this phenomenon:

C1-7 Raphael's *The School of Athens* (colour plate ii)

> The wide versatility of many Renaissance artists—like Alberti, Brunelleschi, Leonardo da Vinci, and Michelangelo—led them to experimentation and to achievement in many of the arts and sciences and gave substance to that concept of the archetypal Renaissance genius—*l'uomo universale*, "the universal man." (p. 524)

One such man was the artist Raphael. In his large oil painting *The School of Athens*, Raphael visually demonstrated the historical connections between the arts and the sciences by juxtaposing a number of great philosophers, artists, and mathematicians of different eras. This was meant to symbolize the new and expansive learning of the Renaissance. Included in his academic fantasy scene were Plato, Aristotle, Pythagoras, Euclid, Leonardo da Vinci, his contemporary hero Michelangelo, and an unobtrusive self-portrait.

In 1983, IBM Canada was involved with the development and sale of an educational packet entitled *I, Leonardo: A Journey of the Mind.* This extraordinary initiative included a video of historical reenactment, a filmstrip, audiotape, time charts, student handouts, suggested projects, and a teacher guide. Also in the kit was an introductory letter addressed, most unusually, to the "Science or Art Department Head," and which included the following paragraph:

> Most recent studies of Leonardo have continued the tradition of focusing on a single aspect of his multi-faceted character: Leonardo the artist, the scientist, the engineer, the city-planner, the architect. "I, Leonardo" attempts to place all these achievements within an interrelated framework; to explain how, for example, Leonardo's observational skills, developed through his work as a painter, extended his scientific insight, and how his scientific investigations enhanced his work as an artist. (1983)

This corporate endeavour not only highlighted the historical connections between mathematics and visual arts, but it also served as a tangible example of how curriculum can be successfully integrated within an interdisciplinary context, using a historical character as a springboard for learning.

The Modern and Postmodern Eras

The last two centuries have seen support for an integrated curriculum wax and wane. A well-rounded European liberal arts education in the 1800s would certainly have involved exposure to art and music as well as to the scientific disciplines (Naisbitt & Aburdene, 1990). While some would argue that mathematics (or the sciences in general) has influenced the visual arts (Dorn, 1994; Golen, 1999), others would maintain that the preponderance of human experience shows the relationship to be vice versa (Shlain, 1991). Dorn (1994) builds a strong case for the arts being heavily influenced by mathematics:

> In art, the twentieth century was a time when: (a) Einstein's discoveries of relativity and of the space-time continuum affected the way space was ordered in painting; (b) Marx's political thought and Freudian psychology radically influenced its form and content; and (c) the positivist ideas of Wittgenstein and the anthropological views of Levi Strauss inspired new uses of language and ritual now seen as the basis for the revisionist and the deconstructivist art of today. (p. 36)

Shlain (1991, p. 19) hypothesizes that "repeatedly throughout history, the artist introduces symbols and icons that in retrospect prove to have been an avant-garde for the thought patterns of a scientific age not yet born." He supports this notion through examples from history, and interprets the major metaphysical blurring between the disciplines as follows:

> While art is thought to be relatively subjective, physics, until this century, scrupulously avoided any mention of the inner thoughts that related to the outer world. Physics concerned itself instead with the objective arena of motion, things, and forces. This stark difference between art and physics blurs in light of the startling revelations put forth by the quantum physicists that emerged from the fusion of the contradictory aspects of light.... Thus "subjectivity," the anathema of all science (and the creative wellspring of all art) had to be admitted into the carefully defended citadel of classical physics. (p. 23)

Still other theorists, such as Vitz and Glimcher (1984), have proposed a "theory of parallelism," in which the advances in both the sciences and the arts are often correlated through simultaneous expressions of perceived reality. They state the following:

> These similarities cannot be treated as accidental because, as it will be shown, often the two works occur at about the same time, and frequently the

> artist's comments make it clear there was influence from visual science or that the artist on his own had discovered the same visual phenomena that contemporary scientists were investigating. Thus, it is argued that the artists' and scientists' parallel conceptual approach to vision frequently resulted in the construction of pictorially similar or even identical works. (p. 37)

Griffiths (2000), in his intriguing paper, "Mathematics at the Turn of the Millennium," describes the dual nature of mathematics, regardless of mutual influence or chronology:

> Indeed, the mathematician G. H. Hardy once said that the practice of mathematics can be justified only as an art form. In fact, there is a parallel with the arts here. Mathematicians, like artists, rely heavily on aesthetics as well as intuition, and it is not uncommon to solve problems while taking a shower or a walk. But with respect to utility, the argument in mathematicians' favor is a strong one. . . . Thus mathematics has a dual nature: it is both an independent discipline valued for precision and intrinsic beauty, and it is a rich source of tools for the world of applications. And the two parts of this duality are intimately connected. (pp. 4–5)

The connections between mathematics and visual art have been nowhere more apparent than in the visible works of art and architecture that combine the two disciplines in both physical and conceptual ways. A list of pertinent artists and architects from the last two centuries would include, among many others, Le Corbusier, Gropius, Escher, Stella, Calder, Moore, LeWitt, Dali, Duchamp, Collins, Mondrian, Silverman, Verhoeff, Colville, and Pollock.

Throughout the course of human history—beginning in ancient Aboriginal cultures, extending through Babylonian, Egyptian, Greek, Roman, Medieval, and Renaissance periods, and continuing on in the modern and postmodern eras—the disciplines of mathematics and visual art have been fused in theory, in educa-

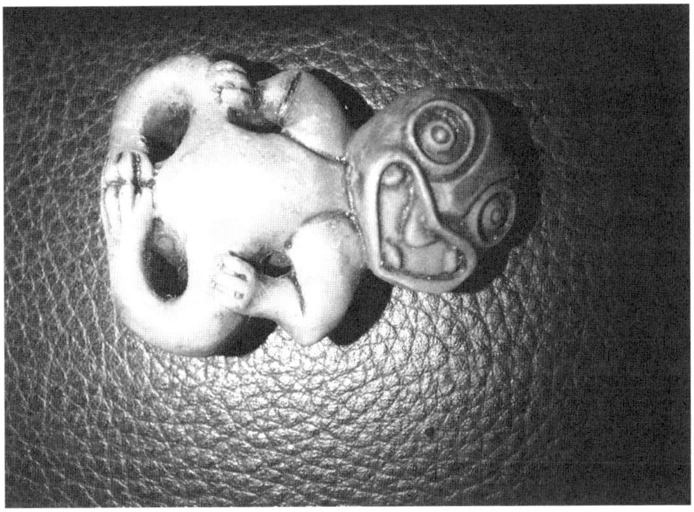

C1-8 Maori tiki

tion, and in the making of images and structures. Integrated learning experiences—whether they are single activities, projects, or extended curriculum—serve to reinforce these chronological connections as students experience an age-old approach to rich learning and application.

Objectives of *Exploring the Math and Art Connection*

The efforts in developing this book were to create/highlight connections between the two traditional disciplines of mathematics and visual arts by way of rich "Learning Experiences." Furthermore, by engaging with this text, teachers should expect that their students will

- Increase their understanding of art education's elements and principles of art/design, or composition, and of mathematical concepts, many of them pertaining to geometry;

- Develop math and art vocabulary, related understanding, and the ability to discuss;
- Develop a more acutely sensitive visual awareness of natural and human-designed objects and environments (e.g., describe, identify, classify, copy, reinvent);
- Create unique, personal artistic statements through drawing, painting, sculpting, assemblage, and digital production using a variety of materials and by applying mathematical concepts;
- Develop a greater overall appreciation of both mathematics and the visual arts;
- Develop the capacity to view mathematical phenomena through an aesthetic/artistic lens, and to view artistic works through a mathematical lens (i.e., different literacies); and
- Improve attitudes towards their abilities in mathematics and in creating works of visual art (i.e., decreasing any existing phobias/anxieties relating to these subjects).

Math and Art Phobia

Math Phobia

> Math phobia is real. It creates a barrier that lies between the student and the learning of mathematics. Early attention must be paid to reluctant and fearful mathematics students. If their fear is not addressed, this barrier grows. My students showed me that a vital key to successful intervention lies in the emotional realm. A teacher must create a safe atmosphere and encourage students to share openly these feelings; all must work together to create this safe place. (Blomfield, 2000, p. 7)

Math phobia or anxiety has been defined as a state of discomfort that occurs in response to situations involving mathematical tasks that are perceived as threatening to self-esteem. It can be experienced as outright fear, intense negative emotions, or other physiological reactions to anything remotely mathematical (Austin, Wadlington & Bitner, 1992). These feelings of anxiety can lead to panic, tension, helplessness, fear, distress, shame, inability to cope, sweaty palms, nervous stomach, difficulty breathing, and loss of ability to concentrate (Posamentier & Stepelman, 1990). Also, according to McCoy (1992), math anxiety is most prevalent among tactile-kinesthetic learners.

The reasons for math anxiety in teachers may include negative math attitudes adopted by their parents and former teachers, inadequate math training backgrounds, and/or lack of general mathematical understanding. Math anxiety may inhibit teachers from acquiring requisite math skills and processes needed to teach children. Research supports the idea that some teachers may transmit their undesirable attitudes and anxiety toward mathematics to their students. Teachers who are ill-prepared or anxious about mathematics tend to use more traditional teaching methods such as lecture, paper-and-pencil worksheets, reliance on rote memory, neglect of developing reasoning or estimation abilities, and the inadvertent creation of a non-participatory classroom. They tend to concentrate on teaching basic skills and concepts with an emphasis on drill-and-practice without necessarily promoting understanding (Fiore, 1999; Selke, 1999).

This is contrary to the current reform-oriented movement toward teaching mathematical concepts and problem solving through cooperative learning and projects (National Council of Teachers of Mathematics, 2000). Elementary school teaching not only requires the teacher to know and understand mathematical content, but a growing body of research has shown that there is a unique and important set of pedagogical skills/thinking referred to as "mathematics knowledge for teaching" (Thames & Ball, 2010), which enables mathematics teachers (i.e., specialists) and teachers of mathematics (i.e., generalists) to teach more effectively by seeing connections between math strands; recognizing common student misconceptions and remediating accordingly; and carefully interpreting multiple solution strategies shared by students in class.

> There is more to real life than rummaging around the kitchen cupboards to find the volume of the Twinkies box or the can of Ravioli. We must distinguish between using things in the world around us to do math upon, and using math to understand the world around us. One is deceitfully artificial, a straw man. The other is transformative, for it encourages people to think, and perhaps to intervene. (Stocker, 2007, p. 48)

Mathematical modelling (representing real-world situations or data using graphs or now easily accessible, powerful software programs) is encouraged as a way of interpreting and representing the world in which the child lives, and of solving related problems that may be of interest to the student or to the entire class.

Art Phobia

Visual art can be viewed as a language that is learned, not unlike reading, speaking, mathematics, music, and movement. Children tend to proceed through developmental stages in visual arts communication. They develop skills in representing their world and expressing their thoughts in two and three dimensions. Children may have different reasons for being interested in visual art (e.g., natural curiosity regarding image making; formal instruction by adults), but all tend to move through stages as they progress physically and intellectually. Unfortunately, many children stop their artistic development at about age ten to twelve years, when the deep desire for realistic drawing ability is not satisfied by their teachers, in terms of classroom instruction. The ability to see, interpret, and create art images is not merely intuitive; rather, it is a learned

Mathema Anathema

Mathema anathema,
A phobia severe!
That vile, loathsome topic,
Which returns to haunt each year.
Mathema anathema,
The c(o)urse one loves to hate!
Its shadow darkens every school;
A nemesis of fate.
Mathema anathema,
Like words of ancient chant;
Resounding through the centuries;
A cry from all who "can't."
Mathema anathema,
Perhaps there be a cure;
If math was simply relevant,
And fun, one could be sure, that...
Mathema anathema,
Would all but disappear,
As interesting lessons,
Now dispel the former fear!

Jarvis 1997

Artifex Ad Artifice

Artifex ad artifice,
The wonder drug of art!
That warm and fuzzy topic,
Which caters to the heart.
Artifex ad artifice,
The course one hates to miss!
Reward for good behaviour,
Is mindless Friday bliss.
Artifex ad artifice,
'Tis sophistry supreme;
The students colour picture books,
And leaf through magazines.
Artifex ad artifice,
Perhaps there be a cure;
If art was simply relevant,
And meaningful, then surely...
Artifex ad artifice,
Would all but disappear,
As synthesis builds competence,
For expectations clear.

Jarvis 2000

ability (Edwards, 1999; Hoffer, 1977). As children enter their pre-adolescent years, they may become critical of the art they create and lack confidence in their own art making. Throughout the developmental stages, children need encouragement and good teaching. It is important that teachers discover where students are positioned, developmentally speaking. They need to support the children in their artistic efforts. The teachers need to become aware of, and avoid, art experiences that may be detrimental (Naested, 2010).

Teachers with art phobia, or anxiety, generally have a poor self-concept relating to their own art ability. They often lack knowledge and understanding of the design elements and principles, art vocabulary, techniques, and the history of art. These teachers may also tend to rely on planned or overly scripted art experiences that may in fact be detrimental to their students' visual art development (e.g., having their students do cut-and-paste activities, complete colour-by-number pictures, or reproduce "craft work").

Why Connect Art and Mathematics?

Distinct, separate subjects of mathematics, science, language arts, music, and visual arts were created artificially by educators, probably for efficiency in curriculum design and teacher planning. But humans do not live in a real world in which subjects are separate. We have perhaps forgotten the natural connections between the subjects. Within the content of this book, we will be stressing teaching strategies and learning experiences that seek to connect visual art and mathematics, and thereby increase student understanding and confidence. These learning experiences are created and presented in such a way as to, wherever possible, honour the integrity of both disciplines, understanding that certain activities may have more of an art or a mathematics emphasis.

Visual Perception in Art and Geometry

Students' visual discrimination and depth perception develop at different times and at different rates, as does their visual-motor coordination. These factors affect the way students interpret and create visual art and two- and three-dimensional geometric images, and may also influence their ability to learn geometry. "If the students are having trouble with visual perception, they will probably experience frustration and failure when trying to recognize pictures and patterns in geometry" (Hoffer, 1977, p. 85). Students who are experiencing difficulty in observing and drawing images of objects need to be guided through observation exercises that have them look more closely to fully grasp the details of the object. This requires careful observation, detail-by-detail, and hand-eye coordination (Jarvis, 2010; Naested, 2010).

> Geometry deals with the properties and relationships of points, lines, angles, surface and solids. There are many concepts in geometry that cannot be recognized or understood, unless the students can visually perceive examples and identify figures and/or properties by associating them with previous experiences. Many geometrical concepts require a visual interpretation as problems are viewed or drawn on paper. (Hoffer, 1977, p. 89)

Depth perception is a developed and learned ability to understand two- and three-dimensional images and objects. It includes figure-ground perception, which is the ability to distinguish the foreground from background, and understanding of objects' position in space, or within spatial relationships. Teachers can plan learning experiences to assist students in developing body orientation though movement in dance, drama, and art. They can develop lessons that require students to learn kinesthetically through constructing, manipulating, carving, drawing, or painting. Visual discrimination is the ability to distinguish similarities and differences between objects. Students learn visual discrimination in relation to colour, shape, size, and thickness by creating objects in two and three dimensions. The learning experiences in this book are designed to assist in the development of both depth perception and visual spatial operations.

Learning Theories that Support an Art and Math Connection

Perhaps the two most exciting areas in contemporary education are human brain research and advances in new technologies. The past two decades have seen tremendous growth and achievements in both of these areas, and these developments have significantly affected education at all levels and continue to do so. We know from brain research that when new knowledge is connected with prior knowledge we tend to understand better, and remember more thoroughly. When we are actively involved in our learning (e.g., constructivism), rather than passive listeners, we benefit from deeper understanding. Furthermore, when we are empowered to select elements of our own learning experiences (choice), the ways in which we learn (differentiated instruction), and to present our new learning to others (communication/teaching), the gains are even more dramatic. We also understand that there are many learning theories that support artistic and mathematical understanding. In the following, we provide short descriptions of a few of these theories.

HEMISPHERIC SPECIALIZATION is a learning theory that suggests that individuals develop a left- or right-brain dominance. The right side of the brain is credited with specializing intuition, spatial relations, and creativity, and the left side is credited with verbal and mathematical ability. In light of this learning theory, connecting art and math experiences will address and develop both sides of the brain (Edwards, 1999).

AUDITORY, VISUAL AND KINESTHETIC LEARNING THEORY was developed by Dunn and Dunn (1978). The auditory learner learns best through words and learning experiences that include talking, group discussion, listening, singing, and listening to music. The visual learner learns through images and learning experiences that include viewing, imagining, reading, drawing, constructing, writing, looking at pictures and creating models or diagrams. The kinesthetic learner learns best through movement and learning experiences that include doing, touching, manipulating, constructing, drawing, and painting. Art and math learning experiences that are connected assist students with these different learning styles.

Howard Gardner (1983) developed the THEORY OF MULTIPLE INTELLIGENCES based on the premise that we have historically viewed intelligence too narrowly. Initially he identified seven intelligences: logical-mathematical, verbal-linguistic, musical, visual-spatial, bodily-kinesthetic, interpersonal and intrapersonal (note: Gardner has subsequently added an eighth intelligence: naturalistic). Visual art and mathematics learning experiences can be developed to address these different types of learning styles.

Visual art experiences develop visual literacy and visual awareness. One important aspect of experiencing and understanding principles of art/design is that students can gain the benefit of more perceptive vision. Likewise, the term mathematical literacy has also become popularized (in preference over "numeracy," which often emphasizes arithmetic/numbers), wherein students become proficient at mathematical skills and problem solving, and grow to view and understand their world with a mathematical perspective.

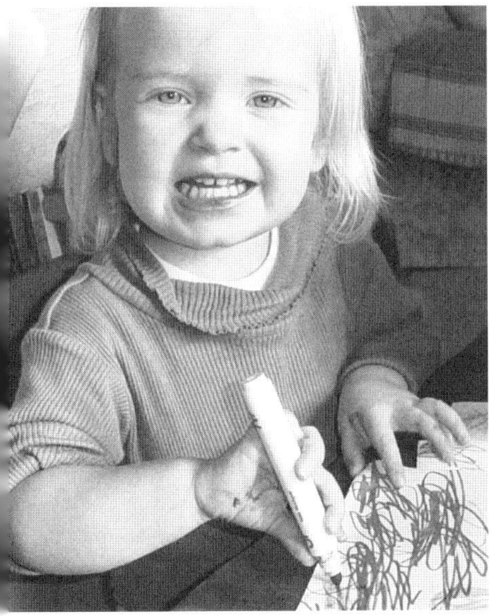

When I was a child,
I played as a child.
When I became a student,
I was encouraged to put away my childish ways,
And decipher meaning from the world through word and number.
When I became a teacher,
I rediscovered the values of play,
And other ways to experience and learn.

Naested, 1991

C1-9 The joy of making lines and images as a child (colour plate ii)

Teaching and Learning Strategies

> Tell me mathematics and I forget; show me mathematics and I may remember, involve me . . . and I understand mathematics. If I understand mathematics, I will be less likely to have math anxiety. And if I become a teacher of mathematics, I can thus begin a cycle that will produce less math anxious students for the generations to come. (Williams, 1988, p. 101)

Williams (1988) paraphrases a Chinese proverb to illustrate what teachers should do to assist students in understanding mathematics—this is the same for understanding visual art and the art process. Teaching strategies in reform-based mathematics classrooms should include small-group collaborative work, the use of manipulatives and visual aids, the use of real-life examples meaningful to the learner, and the development of spatial reasoning through observing, interpreting, understanding, and appreciating our inherently geometric world (National Council of Teachers of Mathematics, 2000). Teachers should facilitate problem-based activities that involve problems and projects that connect mathematics to other subjects and to the "real world" (Jarvis, 2008). Selected theories that assist the teacher in addressing pedagogical approaches that support learning, and that connect visual arts and mathematics, are summarized below. Some of the following names, terms, and definitions have varied over time. Teachers wishing to adopt particular approaches in their pedagogy are encouraged to further research these ideas.

CONSTRUCTIVIST APPROACH: Constructivism is an approach to teaching and learning that encourages learners to take an active role in their learning and construct new knowledge based on prior understanding. "Constructivist classrooms help young people to make sense of their world through actively acquiring, building, and understanding meaning and knowledge in social contexts" (Naested,

Potvin & Waldron, 2004, p. 84). This is closely associated with "active learning theory" and "discovery learning" literature. The premise here is that learners who are actively engaged (cognitively active) in a problem or process are more likely to recall information and/or to solve the problem at hand.

PROBLEM-BASED LEARNING (PBL): PBL was originally developed in the 1970s by a medical professor, Dr. Howard Barrows, at McMaster University in Ontario, Canada, for training medical students using a problem-based approach (Delisle, 1997). PBL features learning that is based on challenging, open-ended problems; often involves students working in collaborative groups; and usually finds teachers assuming the role of "facilitator" of learning, rather than disseminator of knowledge. Students make decisions within a prescribed framework around a problem or challenge, but students often design the process and are responsible for accessing and managing the information they gather. A final product or solution to the problem is evaluated with respect to the process (solution strategies) and the outcome (answer/conclusion). This approach is closely related to "inquiry-based learning" in which learning is based on students' questions (i.e., often a problem of their choosing or their own creation). "Inquiry teaching puts students in the position of researchers, asking and answering questions about information that may range far beyond the boundaries of a single discipline" (Clarke & Agne, 1997, p. 30).

INTERDISCIPLINARY MODEL: The principal aim of integration or interdisciplinary instruction is to present learners with an opportunity to discover relationships that go beyond separate disciplines and that bind together different aspects of our world in some systematic manner (Naested, 2010). This is closely related to "integrative curriculum." Beane (1993) suggests that genuine learning occurs as people "integrate" experiences and insights into their scheme of meanings and the most significant experiences are those tied to exploring questions and concerns that learners may have about themselves and their world.

BRAIN-BASED LEARNING: Neuroscientists such as Jensen, Sywester, Sternberg, Sprenger, D'Arcangelo, and others have written about learning and the brain. According to Jensen (1988), "learning changes the brain because it can rewire itself with each new stimulation, experience, and behavior" (p. 13). The brain adapts and rewires itself. Learning something new, doing something new, listening to something new can stimulate the brain. "The key to getting smarter is growing more synaptic connections between brain cells and not losing existing connections" (p. 15).

DIFFERENTIATED INSTRUCTION (DI): This approach to learning acknowledges the fact that students are at different developmental levels within a given discipline such as mathematics, and that students learn best in different ways. While DI certainly has its roots in frontier one-room schoolhouses in which learning was differentiated (at least by grade level) by necessity, it has been widely popularized by learning theorists such as Carol Ann Tomlinson (1999), who has created an effective model for analyzing factors affecting DI implementation. Canadian Marian Small (2008) has provided the related concepts of "open and parallel tasks" in mathematics instruction, which allow for multiple entry points, multiple versions of problems, and student choice based on their own perceived abilities and readiness.

UNIVERSAL DESIGN FOR LEARNING: This is an educational framework that appears to incorporate or integrate many of the key concepts of the previously mentioned frameworks. The early frameworks for universal design for learning (UDL) focused on students with disabilities or those learning English as a second language, and often included applied special technology. Some ways to integrate the principles of UDL in teaching and learning include using multiple ways of presenting information to address different levels of abilities and learning styles, providing students with choice (content, material, presentation of learning), and incorporating curriculum integration or overlap and continual assessment or feedback.

TEACHING AND LEARNING FOR THE iGENERATION: With the advent of new high-speed Internet technologies in the late twentieth century, access to information, images, open-source software (e.g., Google Art Project, Wikipedia, GeoGebra), and social networking tools (e.g., Facebook, Twitter) have opened up huge new areas and methods for student research and collaboration, making the integration of disciplines such as mathematics and visual arts even more intriguing. Web. 2.0 is a term used to describe how web applications facilitate information sharing, interaction, and collaboration on the world wide web.

Planning Rich Learning Experiences

Over the centuries, a disconnect in public opinion has arguably developed between art and mathematics, causing a commonly perceived disassociation between the two disciplines. While historically the number of artists/mathematicians who well understood the existence and richness of this math/art connection has been relatively small, a growing number of artists and teachers are becoming more aware of this connection. This book will attempt to expand the reader's understanding of both the natural and the built/created connections between visual art and mathematics. Furthermore, it will seek to promote the idea that while visual art images/objects can be interpreted through a mathematical lens; the corollary is also true, in that mathematical or scientific phenomena can be interpreted through an artistic, or aesthetic, lens. In Chapter 1, we contend that as these ways of seeing and interpreting the world are honed and deepened, students can significantly increase their own confidence in their mathematics and visual arts knowledge and abilities, as well as stimulating critical thinking and creativity across all learning domains and pursuits—a pattern that hopefully will continue throughout life.

In Chapter 2, we provide the reader with some general context by highlighting basic vocabulary and concepts traditionally found within the disciplines of mathematics and visual arts. In Chapter 3, we present a series of learning experiences that focus on the math/art connections found within various forms of flora, or naturally occurring plants, trees, and vegetation. Chapter 4 extends this exploration to include learning experiences that focus on the fascinating world of fauna, or living creatures of the land, sea, and skies, that share our planet.

Historically, the natural connections of mathematics and visual art appeared to begin as people developed measurement systems based on the human body and other natural phenomena. Notable artists such as Vitruvius, Michelangelo, da Vinci, Dürer, Rembrandt, and Giacometti developed measurement systems to describe lengths and quantities by using body parts, such as fingers and hands, and using body movements to measure pacing. The understanding of the human body and appendages, along with the related study of proportion, greatly assisted the artist and the engineer in working with natural and human-designed objects and environments such as those created by architects like Ictinus, Callicrates, Phidias, and Le Corbusier. Chapter 5 examines mathematics and visual arts learning experiences relating to the human figure, and Chapter 6 features activities based on human-designed environments such as buildings and cities.

In Chapter 7, we turn our attention to learning experiences that involve connections between mathematics and visual art that relate to designed objects and games. Finally, in Chapter 8, we discuss how curricular integration can be implemented within a single lesson, an extended project, or a unit of study. We further provide a set of helpful appendices that comprise a glossary of key mathematical terms, a glossary of key visual arts terms, two sample extended projects, and a list of other related print and online resources.

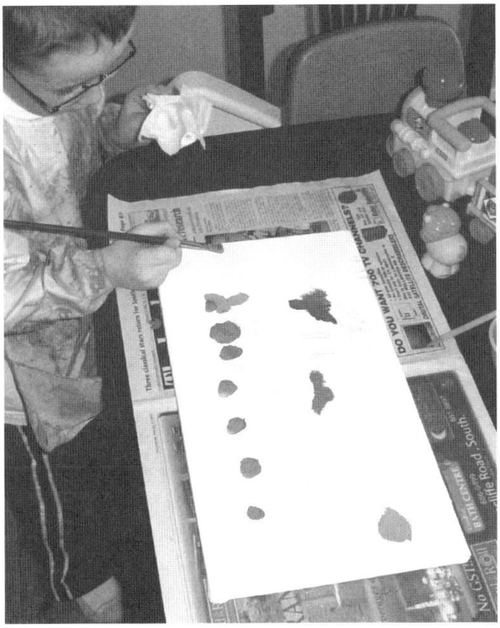

C1-10A Drawing the train

C1-10B Painting the train (colour plate ii)

C1-10C Contextualizing the train

2 Understanding the Language of Mathematics and Visual Arts

Introduction

To describe any discipline, be it art, math, drama, language arts, dance, science, and so forth, academics have developed key concepts and descriptors for what might be considered elements and principles to categorize phenomena. These categorizations assist in developing understanding, communicating with others, and creating new theories of understanding within the discipline. Most disciplines or subjects are taught separately, which often has a fragmentary effect on learning and related understanding. Mathematics and visual art are not polar opposites, as they are often portrayed, but rather can be viewed as complementary ways of understanding and communicating. Artists and mathematicians who understand the richness of both mathematics and visual art understand the "language" and culture of both disciplines, and thereby become aware of the natural connections between the two. "Content, knowledge and understanding can be drawn from one discipline and used to enrich and apply to another . . . with a holistic approach teachers can draw content from one discipline and use it to enrich another, rather than stripping down any one subject" (Naested, 2010, p. 4).

In this second chapter we will present two sections dealing with "understanding the language" of a discipline, namely, mathematics and visual arts. We will highlight rich examples of problem-based learning in mathematics and learning experiences/activities in visual art, also providing the reader with a sample art assessment tool. Finally, we will make the case that while mathematical objects can be viewed through an aesthetic/artistic lens, the corollary is also true—that visual artworks can be viewed through a mathematical lens. These lenses provide students with rich ways of seeing and interpreting their immediate environments and the larger world, and hence improving learning and creative capacity.

Understanding the Language of Mathematics

What is mathematics? How is it different from science, physics, chemistry, or biology? Carl Friedrich Gauss (1777–1855), considered the "Prince of Mathematicians," referred to mathematics as the "Queen of the Sciences," emphasizing its foundational role in all of the other modern sciences mentioned above. Small (2008, pp. 5–7) describes four distinct perspectives held by individuals regarding what they perceive mathematics to actually be: (a) a set of procedures, (b) a hierarchy of concepts and skills, (c) a study of pattern, and (d) a way of thinking. How one views mathematics ultimately determines how one teaches the subject, including the ways in which learning experiences are planned, implemented, and assessed in classrooms/courses.

Mathematics consists of several major branches which, although often overlapping in various ways, have been recognized as areas of related skills or concepts, and upon which much of modern curriculum is now based. For example, in Ontario, Canada, the provincial curriculum features five major mathematical "strands" that are consistent throughout the K–8 documents:

- Number sense and numeration
- Measurement
- Geometry and spatial sense

- Patterning and algebra
- Data management and probability

Similar categorizations can be found in the Western and Northern Canadian Protocol (WNCP), the Quebec curriculum, and the Atlantic provinces curriculum (APEF/CAMET) (Small, 2008, p. 8).

Beyond the content areas, there has also been a recent emphasis in the United States and Canada on what are described as the "mathematical process" expectations. Again, while these differ somewhat in name and number within the four main education curricula in Canada, these math processes, which are to be fostered among students and assessed throughout all mathematics learning, include elements such as

- Problem solving
- Reasoning and proving
- Reflecting
- Selecting tools and computational strategies
- Connecting
- Representing
- Communicating
- Practicing mental mathematics and estimation
- Using Technology
- Visualizing

Mathematical process expectations can most effectively be addressed inside a classroom in which problem-based learning experiences are taking place on some form of regular basis (Jarvis, 2008).

In an era of "reform-based" mathematics education (National Council of Teachers of Mathematics, 2000; Ontario Ministry of Education, 2005), in which key elements such as problem-based learning (PBL), cooperative group work, manipulatives, technology, and varied assessment are being emphasized, it is not difficult to understand why teaching mathematics remains a complex undertaking. Jarvis and Franks (2011, p. 28) indicate that twenty-first century mathematics educators must possess at least four overlapping areas of specialized knowledge and related competencies:

- They must have a thorough understanding of the **mathematics content** (e.g., Ontario curriculum expectations as listed/described within the five mathematics strands);
- They must also master, and be able to implement with confidence, an ever-expanding range of **mathematics resources for teaching** such as manipulatives (e.g., linking cubes, geoboards, pattern blocks, algebra tiles) and technologies (e.g., graphing/CAS calculators, computer software, Internet web resources, interactive whiteboards);
- They must develop what Thames and Ball (2010) describe as unique "**mathematics knowledge for teaching**," or a pedagogical awareness of how mathematical topics are connected, and how students acquire, organize, and communicate their mathematical thinking (including an awareness of common student misconceptions and alternative solution strategies); and
- They must incorporate **balanced instructional planning**, featuring both the more traditional methods of teacher-directed lessons and skill-based practice (e.g., notes, textbook questions), combined with regular opportunities for rich, interactive problem-based learning.

C2-1 Manipulatives and problem-based learning in mathematics education
(colour plate iii)

Clearly, as we shall also see with visual art education in the next section, mathematics education demands much from a teacher both in terms of the mastery of content, and in the development of effective pedagogical practices.

Understanding the Language of Visual Art

What is visual art? How is this different from craft? There has been much debate and much written about what the definition of art actually is. Some would argue that art is primarily personal expression. "To express in a creative, innovative manner, one must be aware of one's inner and outer world. A creative expression reflects the artist's experiences, observations and understanding" (Naested, 2010, p. 21).

It is often *not* art being experienced in classrooms when

- A step-by-step, craft-making approach is presented by the teacher and followed by students;
- Stereotypical images, photocopies, clip art, and pre-drawn images are used; and
- Art understanding is not developed or discussed during and/or following the experience.

We contend that it is much more likely to be a *meaningful art learning experience* when

- Individual or personal choices are made by students themselves regarding the elements and principles of art/design in both the planning and creation stages of works of art;

- Affordances/restrictions of different media are considered in selecting tools for projects; and
- Art understanding is developed and communicated to one's peers and/or to other audiences.

Art making is jealous, for it requires the whole person!
—Author Unknown

Turning an Activity into an Art-Learning Experience

The following are instructions for a "four-fold fortune teller." This four-fold origami exercise can be made into many game variations of question and answer. By following the step-by-step instructions to create the four-fold paper object, one does not create an art object. However, this four-fold object can become a form of artistic expression if certain conditions are met, which move the experience beyond just following the directions step-by-step. As you read through these instructions, think about how this "activity" could become an "art learning experience."

Developing the Four-Fold Fortune Teller Activity into an Art Learning Experience

C2-2 Four-fold object design
(colour plate iii)

We suggest that there are several ways to move the four-fold fortune-teller activity towards a richer art-learning experience. The object itself could be decorated by the student, consciously incorporating one or more of the elements or principles of art/composition/design as described below. The object could be decorated using images and colours representing one or more art styles from art history. The visible and hidden

STEP-BY-STEP INSTRUCTIONS FOR THE FOUR-FOLD FORTUNE TELLER

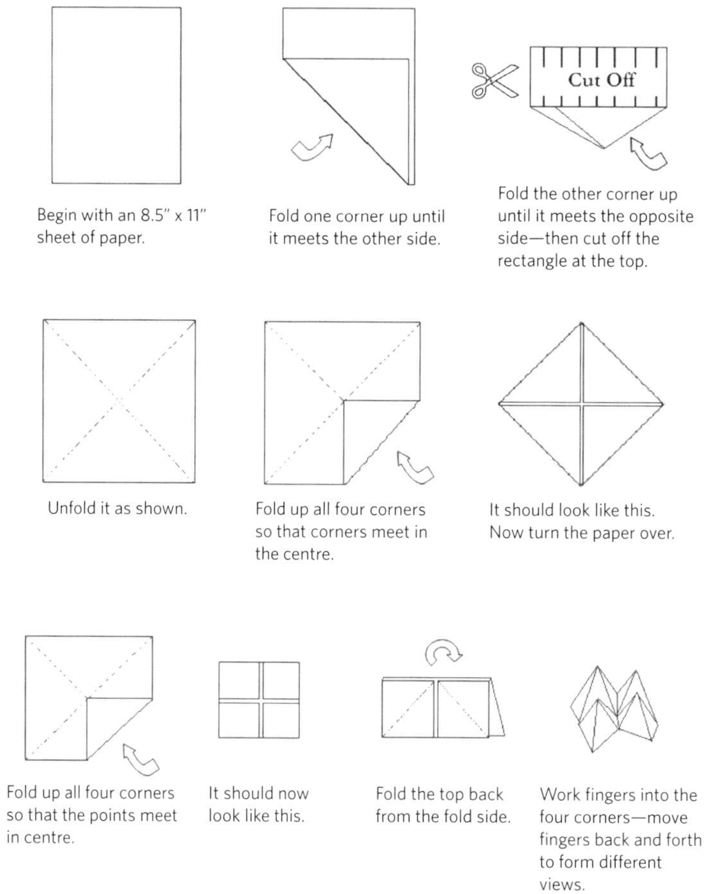

content of the flaps could include the names of random objects, places, people, or themes that, when selected and matched up, could form the basis of a creative project which could draw upon several disciplines (integrated). Students could explore whether other folding patterns with a piece of paper could be developed that would produce more than four "secret" flaps, or analyze the mathematical probability of selecting any one flap. What are other ways to modify the step-by-step activity so that it involves more student creativity?

Elements of Art/Design

The following is a brief description of the elements and principles of art/design, and other art terms that will assist the reader in the discussion and development of art learning experiences. Composition in the visual arts is the way in which elements and principles are visible in a work of art. Although the number and names of the elements often vary depending on the resource, six elements of art/design are commonly described as the building blocks of a composition or visual statement: line, shape, texture, space (form), value (tone), and colour. A visual composition (two dimensional, three dimensional, or relief) will have within it one or more of these six elements.

Line: The visual path (straight or curved) left by a point moving through space. Lines are found in both natural and constructed objects; lines define space or are used to create decoration, pattern, or texture.

Shape: The external outline or form of a two-dimensional object. Shapes can be geometric (often involving straight lines) or organic (often involving curved lines).

Texture: The feel or appearance of a surface or substance.

C2-3 Line (colour plate iii)

C2-4 Shape (colour plate iii)

C2-5 Texture (colour plate iv)

Space (Form): the area around, within, or between images (three-dimensional space created visually through techniques such as overlap, placement on the picture plane, and amount of detail).

Value (Tone): The property of a colour that describes its degree of lightness or darkness. Adding black pigment darkens or lowers the tonal value of a colour (its shade), whereas adding white pigment lightens or raises the tonal value of a colour (its tint).

Colour: The range of different wavelengths of light seen by the human eye when an object reflects or emits light. Its four characteristics are hue, value, intensity, and temperature.

C2-6 Space (Form) (colour plate iv)

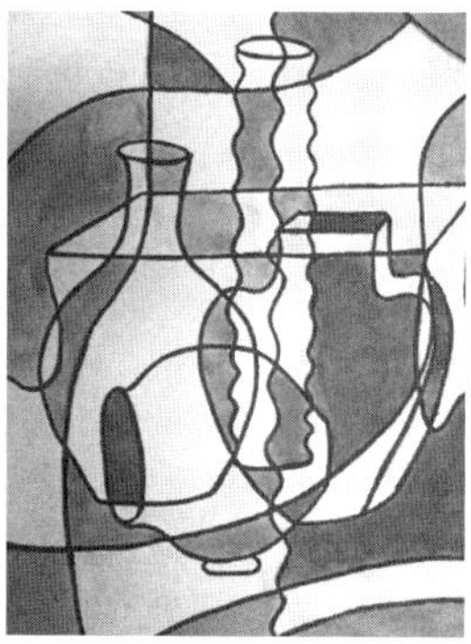

C2-7 Value (Tone) (colour plate iv)

C2-8 Colour (colour plate v)

Principles of Art/Design

If the elements can be thought of as the basic "tools in the artist's toolkit," then the principles can be viewed as the ways in which these tools are put to use in projects, or more specifically, in the creation of images and sculptures. Again, while the number and names may vary between different art resources, the principles generally include the following seven items: balance, contrast (variety), emphasis (dominance), pattern (repetition), rhythm (movement), proportion, and unity (harmony).

Balance: Achieved when all parts of an artwork are arranged either formally or informally around the axis. Formal balance occurs when equal or very similar elements are placed on opposite sides of the axis. Informal balance occurs when the elements on either side of the axis are, in an overall sense, visually equal in size (e.g., one large object on one side, two medium-sized objects on the other side). For bilateral symmetry, objects match on either side of a given axis; for approximate symmetry, there are some differences.

C2-9 Balance (colour plate iv)

C2-10 Contrast (Variety) (colour plate v)

Contrast (Variety): Ways of adding interest to a composition. Contrast places emphasis on an area and keeps the image from being visually "boring." Varying the elements of composition will keep the image from having too much unity/harmony.

Emphasis (Dominance): Focusing attention on one area of an artwork makes it dominant. Think of emphasis as a focal point that draws the eye first, after which the rest of the object is viewed. Emphasis controls the sequence in which we view an art object. Another method of using emphasis is when a dominant element is so strong that the rest of the elements are subordinate (e.g., if a painting is saturated in the colour blue, it affects the viewing of the painting as a whole).

C2-11 Emphasis (Dominance) (colour plate v)

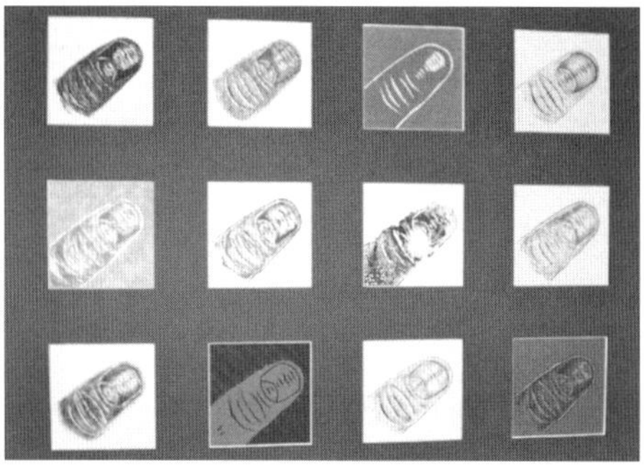

C2-12 Pattern (Repetition) (colour plate vi)

C2-13 Rhythm (Movement) (colour plate vi)

Pattern (Repetition): A visual repetition made from one or more elements of composition (e.g., shape, colour, value, line, or texture), or involving the repetition of symbols/ideas/motifs. Elements can be repeated in a regular or irregular fashion.

Rhythm (Movement): Expresses movement by the repetition of one or more of the elements of art/design, often in close proximity within the work of art. Rhythm produces a perception of movement because the viewer's eyes are purposefully being lead by the artist from one place to another within a work of art.

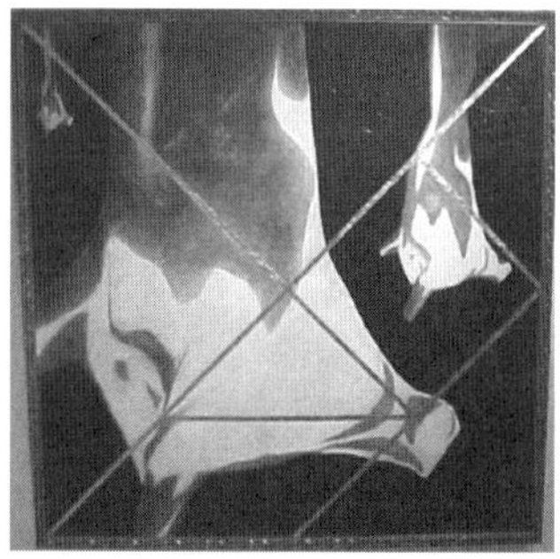

C2-14 Proportion (colour plate vi)

Proportion: Proportion refers to the relative size and scale of the various elements in a design, where the critical issue is the relationship between objects, or parts, of a whole.

Unity (Harmony): Depending on their intent, artists will adjust the parts of an artwork in such as way that they relate to each other in a complementary/unified fashion, or quite the opposite situation, where they purposefully create disunity and tension in the work.

C2-15 Unity (Harmony) (colour plate vii)

Visual Art Media

In order to further understand visual art, it is helpful to understand the use of the various media and materials that are commonly used. The following are many of the terms that have been categorized under various media headings: printmaking, fabric arts, painting media, painting techniques, drawing techniques and clay techniques. Throughout this book, reference is made to these terms. At the end of the book, an appendix of key visual arts terms can be found for further reference.

C2-16 Printmaking with stamp and colour ink (colour plate vii)

Printmaking: block printing; brayer; collage; collograph; embossed printing; etching or intaglio; monoprint; planographic or lithographic printing; relief print; rubbings or frottages; stamp print; stencil, serigraphy or silk-screen printing; transfer printing

Fabric arts: appliqué, batik, collage, fabric paint and crayons, macramé, pulling strings, quilting and stuffing, stamping fabrics, stencil and screen printing, stitchery, tie-dye, weaving (warp and weft)

Painting media: acrylic, chalk, chalk pastels, collage, crayons, gouache, tempera paint (blocks, powder, liquid), oil pastels, oil paint, watercolour

C2-17 Watercolour painting (colour plate vii)

Painting techniques: chalk painting, collage, crushed-paper batik, dry brush, fingerpainting, paint and string drip, splatter painting, straw or blow painting, wash, wax resist, wax rubbing, wet on wet, etching, tissue painting, tissue transfer painting

Drawing techniques: blind contour, contour, cross-hatching, gesture, smudging, ghost

Clay techniques: coil, slab, modelling, carving, relief, pinch-pot, armatures, figures, firing, glazing

C2-18A Clay works (bisque ware) after kiln firing

C2-18B Kiln oven for clay firing
(colour plate viii)

Planning Learning Experiences in Math and Art

There are many forms of lesson plans available; the main components and terms vary, but the main elements that should not be left out are the *why*, the *what* and the *how* of the learning experience. Why are students involved in the learning experience, and what are the expected learning outcomes or curricular expectations? How are students going to be drawn in, intrigued, and engaged, or how is the new learning going to connect to their prior learning? What instructional strategies will be incorporated into the lesson? What resources and art materials are needed? What process or tool will be used to assess the learning experience? The following is one suggested lesson plan format for designing rich learning experiences:

LESSON PLAN FORMAT

Topic/Title/Grade Level:

Purpose/Learning Outcomes (the *why* of the lesson)

- What will students learn in this lesson?
- Description of a learning outcome (e.g., Students will be able to . . . ; Students will develop an understanding of . . . ; Students will develop an awareness of . . .)

Motivational Strategies (the *what* or substance of the lesson)

- How will students' interest be captured and their engagement be kept high throughout?
- How will the lesson be connected to what students already know or are involved in (examples, demonstration, previous research on the topic, guest speaker, etc.)?

Materials/Equipment (the *resources* to be prepared before the lesson starts)

Process (the *how* of the lesson)

- Introduction: Review, motivate, state objectives/intended outcomes, set the scene, discuss new terminology, briefly describe the lesson
- Body: Sequenced activities, descriptions of instructions, possible demonstration, grouping arrangements of students, specialized material/safety procedures
- Closure: Review, summarize, synthesize learning, share work, comment on successes/challenges, connect to the lesson, assign homework, cleanup
- Assessment: Describe how/when you will assess the attainment of objectives (e.g., checklist, journals, anecdotal records, quiz, teacher observation, class discussion, student demonstration, notebooks, collected work, question and answer period, rubric)

Extensions: Other media/materials (e.g., drawing transitions into a related painting)

Connections: Other subject areas (e.g., art with mathematics being integrated)

Assessing Art Learning Experiences

Your assessment process can serve as a guide to help students toward self-evaluation, to consider others' ideas, to weigh opinions, and to increase understanding of art and the use of art materials. (Naested, 2010, p. 75)

Teachers often find assessment of students' artworks to be a relatively difficult, cumbersome, or subjective process. In most well-planned art learning experiences, students are highly engaged in the learning as a function of the personalization and creativity that are purposefully built into the lesson/project. This kind of engagement with learning allows the instructor ample time and opportunity to observe students during the planning, implementation, and communication stages of the lesson/project, making assessment less intimidating (and often quite enjoyable) for classroom teachers.

The following is an adaptation of a rubric (i.e., the text that would appear in the rubric matrix) developed by Tonya Seyer, instructor in art education at Mount Royal University, in Calgary, Alberta, for the assessment of a mandala. We have adapted the rubric to include the assessment of mathematical concepts. Learning experiences for mandalas are found in chapter 7.

The Mandala

This project will be assessed on the following criteria:

Quality of Symbols

5 Symbols are unique and accessible; thought is evident in choice of colour

4 Symbols are innovative, but standard (common) colour scheme used, or vice versa

3 Symbols and colours are both standard (common)

2 Symbols are randomly put together; little consideration paid to colour

1 Work is visually illegible

Quality of Pattern

5 Pattern is unique and symbols are innovatively put together

4 Positive and negative space are equally balanced and are visually interesting

3 Positive/negative are equally balanced, but one contains more detail than the other

2 Pattern is standard (common); positive and negative spaces are unbalanced

1 Pattern is random and unintelligible

Technical Skill

5 Graphics are rendered with control and attention to detail; work is well presented

3 Graphics are rendered with some degree of control and attention to detail

1 Graphics are rendered with little/no degree of control and attention to detail

Mathematical Concepts

5 Demonstrates a full understanding of rotational symmetry (transformational geometry)

3 Demonstrates partial understanding of rotational symmetry

1 Demonstrates little/no understanding of rotational symmetry

Overall Effect

5 Visually captivating; sustained viewer engagement with the artwork

3 Visually interesting; viewer engagement with the artwork

1 Visually banal; little/no viewer engagement with the artwork

Looking at Art Through the Artist's Lens

Feldman's Four Steps of Art Criticism

Feldman (1967) developed a method of assisting viewers and artists in discussing and appreciating works of art. In applying his method, the viewer developed observational abilities, was encouraged to connect the visual experience to prior life experiences, and deepened understanding of art images. To encourage sustained observation and deeper understanding of artworks, Feldman describes a four-step procedure: description, analysis, interpretation, and judgment.

1. Description (The Details): Focus on media used, dimensions, location, date, recognizable objects/forms in the artwork; What are the viewer's first impressions of the work of art?
2. Analysis (The Design): Focus on the organization of the artwork according to the elements of art/design (e.g., line, shape, texture, space, value, colour) and the principles of art/design (e.g., balance, contrast, emphasis, pattern, repetition, rhythm, unity); What kind of balance is used? Where is the focal point, i.e., what is the first thing seen? What kind of illusion of space is used (deep, shallow, flat)?
3. Interpretation (The Meaning): Focus on the intended and/or perceived purpose or meaning of the artwork; How does this work compare to other works done by the artist? How does the artwork make you feel? What ideas are communicated directly and indirectly (intimated via symbols/visual metaphors)?
4. Judgment (The Evaluation): Focus on the viewer's personal evaluation of the artwork based on the detailed observations and research from the preceding three steps; In the opinion of the viewer, does the artwork attempt to imitate nature, express feeling or emotion, or raise awareness of a pri-

vate or social issue? Is the artist successful in achieving these perceived purposes?

Looking at Art Through a Mathematical Lens

Feldman's (1967) method of assisting viewers and artists in observing, discussing, and appreciating works of art can be further developed to view visual art through a mathematical lens, i.e., learning between the lines. For example, during the analysis stage, mathematical concepts such as ratio, proportion, length, or transformations can also be identified in the work of art. During the interpretation stage, one should consider whether these mathematical symbols/proportions/processes are intentional or coincidental. In the judgment stage, one might ask, How do these mathematical concepts serve to enhance the visual statement?

This theme of looking at art through a mathematical lens will be further developed throughout this book, particularly in terms of rich learning experiences and related vocabulary. Furthermore, we will also look at mathematical objects (e.g., functions, graphs, shapes, three-dimensional models) with an artistic/aesthetic lens. These lenses—the artistic/aesthetic and the mathematical—when more fully developed, may allow individuals to better interpret and describe their surroundings, as well as increase their own creativity and insight.

Summary

In this chapter, we have explored what it is to understand the language of both mathematics and visual art. Key components and vocabulary from both domains have been presented in brief, such as the elements and principals of art/design, broad categories of visual art media, and basic mathematical content strands and processes. We have discussed what it is that makes an art activity a rich learning experience, and have demonstrated this through the modification of an activity that we consider limited in scope. We have provided a sample lesson plan for an art activity along with a related assessment tool. Feldman's four-step process for art criticism, which has been widely used in North American schools for the past four decades, was described in some detail. Finally, we suggested that artworks can be viewed through both an artist's lens and a mathematical lens, a concept that will be further developed throughout the remaining chapters.

C2-19 Two lenses on the world

3 Mathematics and Art in Flora and Landscapes

Proportion is a principle in Nature which is a purely mathematical one and to be rightly interpreted by man through the means of geometry; therefore geometry is not only the gateway to science but it is also a noble portal opening wide into the realms of art. (Colman & Coan, 1971, p. 1)

Introduction

Artists and mathematicians throughout history have been inspired by the form, shape, pattern, colour, and beauty of the many types of flora found on the planet. Humans have invented a formal system to recognize and classify all types of plants, trees, and other vegetation. The formation of various flora is dependent on what the ancient Greeks considered as the four basic environmental elements of earth (e.g., the soil type), air (e.g., wind patterns), water (e.g., moisture from rain and snow), and fire (e.g., the sun). Yet there is an incredible order in the way in which flora develop that can be interpreted through the lenses of geometry, science, and the visual arts. Readers are encouraged to observe patterns, mirrored images, and geometric shapes in the natural environment. Art and math experiences in this chapter have been developed to stress observation skills and to encourage the development of drawing, painting, and sculpture abilities, along with mathematical understanding.

C3-1 Exploring the natural world (colour plate viii)

Formal systems to recognize and classify both flora (plants) and fauna (animals) have been developed. In science, a formalized, taxonomic system of naming plants and animals using binomial nomenclature (a two-part Latin name for the genera and species of each known plant and animal) is widely used. For example, the Monarch butterfly is referred to as *Danaus plexippus*. This system was first introduced by a Swede, Carl von Linné, in an attempt to end the ambiguity of vernacular or common names that often differed depending on location and tradition. His nomenclature system was applied to plants in 1753 and to animals in 1756.

There are other ways of describing flora and fauna. For example, mathematics can be viewed as another language—involving numbers, symbols, and a specialized vocabulary—for documenting the natural world (forms, shapes, patterns). Some have also referred to visual awareness, appreciation, or the act of being an artist as yet another form of observing and documenting the natural world. Frank (1993) claims that what he has not drawn, he has never truly seen. Artists are aware of the natural environment, especially the smaller details that often go unnoticed (shapes and lines, colours and tones, and textures).

Artists observe closely in order to understand, interpret and reinterpret, or communicate to others what they have seen or felt. Mathematicians use the same methods to develop spatial reasoning and to mathematically reinterpret the natural world. Close observation often leads to a much deeper appreciation of the observed object. In order to really understand the natural world, one must integrate new observations with one's previously understood concepts and assumptions. It is then that one can reinterpret, restate, and attempt to communicate this new understanding to others (visually, verbally, musically, kinesthetically, and mathematically).

C3-2 Corn cob patterns (colour plate viii)

It is important that young learners are given many opportunities to observe their natural world, and whenever possible, to bring the natural world into the classroom and other spaces. Various mathematical and artistic experiences presented in this chapter require that learners are able to see and touch and to really observe natural forms in order to gain greater awareness and understanding. Some visual information can also be gained through pictures and videos, from books, the library and on the Internet.

Appreciating Plants

> *I developed my love of flowers. Their shapes came folded inside tiny seeds, so small that fifty of them filled the bottom of a flat seed packet. They were miraculously encoded from the beginning, little specs of dark matter that we shook into our hands, then sowed into flower beds. They sprouted, then opened out in a studied and careful program of increments. Now that was astonishing, all those compressed unfoldings and burstings, but no one said so. No one made a fuss when the seeds actually performed: sprouts, leaves, the long rivery stems, and finally the intricacy of blossoms.* (Shields, 2003, pp. 147–148)

Plants are an inseparable part of our ecosystem. Without plants, neither humans nor animals could survive. Plants provide us with food, oxygen, building materials, paper and a host of other important and useful things. Plants are useful, valuable, interesting, and beautiful (Gates, 1989). We need to encourage students to develop an appreciation for flora.

C3-3 Examining flowers (colour plate viii)

LEARNING EXPERIENCE

Observing and Documenting Plant Growth

- Have students plant seeds from flowers, herbs, fruits, or trees, then water them, observe their growth, and measure, draw, document, paint, and illustrate this development in a calendar, "growth chart," or graph. Students could even create their own flowerpots from clay.

Plant	Date:	Date:	Date:	Date:	Date:	Date:
Plant #1						
Plant #2						

C3-4 Observing/recording tomato plant growth (colour plate ix)

Patterns

He was learning a lot of new things. Some of them were things that he had already experienced, and weren't really new, but that he had never perceived before. And he hadn't perceived them because he had be-

> *come accustomed to them. He realized: If I can learn to understand this language without words, I can learn to understand the world.* (Coelho, 1993, pp. 43–44)

A pattern is a visual repetition made from one or more elements of art/design (line, colour, shape, texture, form, tone). The elements are repeated in a regular or irregular fashion. The human mind has developed a formal system of thought for recognizing, classifying, and explaining patterns. "By using mathematics to organize and systematize our ideas about patterns, we have discovered a great secret: nature's patterns are not just there to be admired, they are vital clues to the rules that govern natural processes." (Stewart, 1995, p. 1). Douady and Couder (1992), two French mathematicians, demonstrated that the mathematical patterns in plants arise from universal laws of the physical world (e.g., phyllotaxis, recursion patterns). For these reasons, artists, naturalists, and mathematicians have all found flowers fascinating.

LEARNING EXPERIENCE
Printmaking and Wax Rubbings with Plants

- Students can use plant parts as stamp prints as a printmaking method. Ink the plant parts and stamp onto paper to create pattern. Discuss and show examples of the concept of patterns and how they are created. Patterns can be regular or irregular, symmetrical or asymmetrical.
- Wax rubbings of plant parts can be done by placing the plant part under lightweight paper and rubbing over with crayons. Again, patterns can be created from these. Students can overlap one rubbing to create the illusion of depth or to create a centre of interest. Other elements and principles of composition can be investigated or emphasized when creating artworks by combining images using rubbings (watercolour wash technique over wax crayons which resist the paint).

- Plant parts can be drawn, simplified, and stylized for a relief print stamp. A relief print can be created by cutting cardboard shapes and pasting them on another piece of cardboard. Ink this cardboard stamp and print it on paper. This decorated paper can be used as wrapping paper.

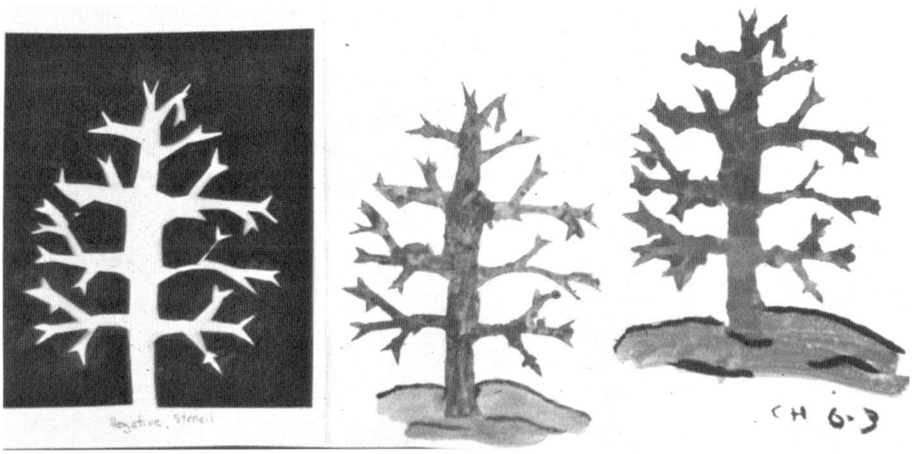

C3-5 Relief stamp print of tree (colour plate ix)

Golden Ratio $$\frac{1 \pm \sqrt{5}}{2} \approx 1.618033988749895$$

Many plants and trees display the pattern of a golden ratio which is approximately 1.6. The golden ratio (1:1.6), which is also known as the golden proportion, phi proportion, divine proportion, golden rectangle, golden section, or golden mean, can be found in many flowers, seed heads, petals, and leaves. For example, when an apple is cut in half horizontally, one notices a perfect star-shaped pattern of the five cavities. Because the points of the star, when connected, form a regular pentagon (equal sides/angles of 72°), this shape contains many golden ratios. The individual florets of the daisy (and of a sunflower, as well) grow in two spirals extending out from the centre. The first spiral has 21 arms, while the other has 34. These are Fibonacci numbers (e.g., 1, 1, 2, 3, 5, 8, 13, 21, 34, etc.) and

C3-6 Apple with star pattern and golden ratio (colour plate ix)

their ratio (when divided) is the golden ratio. The spirals of a pine cone from its centre have 5 and 8 arms, respectively (or 8 and 13, depending on the size), which are also Fibonacci numbers. Likewise, a pineapple has three arms of 5, 8, and 13. Scientists speculate that plants that grow in spiral formation do so in Fibonacci numbers because this arrangement makes for the perfect spacing for maximum growth potential, thereby increases the chance of survival of the plant.

The golden ratio provides flora with not only aesthetic properties, which the viewer and artist can appreciate, but also photosynthetic properties, as scientists tell us that these design arrangements (i.e., phyllotaxis) allow for the best arrangement of petals to maximize moisture absorption and light exposure. This ensures that each leaf on a plant gets maximum exposure to the sun, for example, and that falling rain is captured and redirected towards the stem, and hence the roots, of the plant.

Fibonacci Series

Why is the four-leaf clover so rare, and thought to be lucky? Could it have something to do with the number four not being a Fibonacci number? "A very curious pattern indeed occurs in the petals of flowers. In nearly all flowers, the number of petals is one of the numbers that occur in the strange sequence" (Stewart, 1995, p. 4). The Fibonacci number pattern was documented centuries ago and has been studied ever since (Stewart, 1995).

The Fibonacci sequence, named after the famous twelfth-century Italian mathematician, Leonardo Fibonacci, who first recognized the unique pattern, is a special sequence of numbers in which each is the sum of the two preceding numbers (e.g., 1, 1, 2, 3, 5, 8, 13, 21, 34, 55, 89 ...). Furthermore, the ratio of each successive pair of numbers in the series approximates phi (1.61803 ...), as 5 divided by 3 is 1.666, and 8 divided by 5 is 1.60, etc. Fibonacci numbers are reflected in the number of petals of flowers; for example, lilies and irises have 3 petals, wild rose and buttercups have 5, delphiniums have 8, marigolds have 13, asters have 21, and daisies have 34, 55, or 89.

C3-7 Daisy (colour plate ix)

LEARNING EXPERIENCE

Observing and Documenting Plant Characteristics

- Have students find and observe flowers and other plants (i.e., field trips to local gardens, greenhouses, or forests; gardening magazines; library or Internet research) that have mathematical characteristics. They should draw each plant in order to illustrate these characteristics. This learning experience can be broken down into smaller segments by having students consider one characteristic or term at a time. See the Appendix of Key Mathematics Terms for descriptors of each characteristic. Students might use a table on a large sheet of drawing paper to record their results.

Characteristic	Plant 1:	Plant 2:	Plant 3:	Plant 4:	Plant 5:	Plant 6:	Plant 7:
Circle							
Spiral							
Pentagon							
Symmetry							
Radial Symmetry							
Golden Ratio & Fibonacci Numbers							

C3-8 Romanesco with radial symmetry pattern
(colour plate x)

C3-9 Fiddleheads with spiral growth pattern
(colour plate x)

Spirals: The Calm at the Centre

> *If you watch a daisy flower grow, over time you'll see its central yellow florets ("little flowers") emerge mysteriously from the living eye of the plant and grow, whirling outward more slowly toward the periphery. In fact, the name daisy derives from "day's eye" the eye of the day. (Schneider, 1994, p. 147)*

The spiral is a powerful visual form found throughout the universe, in the galaxies, and in smaller natural forms, such as flowers, animal horns, and seashells. From the earliest record of visual statements, artists and designers have used the spiral form in their work (Hubbard, 1986). The spiral has also been seen as the expression of mechanical laws that govern phyllotaxis, which is the study of the distribution of leaves in plants. "The preferred angle of 137.5 degrees allows for the optimal packing of primordia placed sequentially along a spiral" (Ball, 1999, p. 106). Fibonacci numbers and spiral patterns appear in branching plants, flower petals and seeds, and leaf arrangements, as well as on pineapples, and in apples and pine cones. The number of petals or seeds on a flower, or scales on a pine cone, that occur within spiralling growth patterns are often members of the Fibonacci number sequence (e.g., 3, 5, 8, 13, 21, 34). In flowers, for example, we rarely find four-petal plants existing in nature. Rather, we find instances of 1 petal (white calla lily), 2 petals (euphorbia), 3 petals (trillium), 5 petals—which are the most common (columbine), 8 petals (bloodroot), 13 petals (black-eyed Susan), 21 petals (Shasta daisy), and 34 petals (field daisy). When one observes the heads of sunflowers, one notices the seeds forming two series of gradual curves, one winding in one direction and one in another—the number of spirals in pairs of either 21 and 34, 34 and 55, 55 and 89, or 89 and 144. In pine cones, there are either 8 spirals from one side and 13 from the other, or 5 spirals from one side and 8 from the other. And in a pineapple, the individual outer segments are also found to be 8 in one direction and 13 in

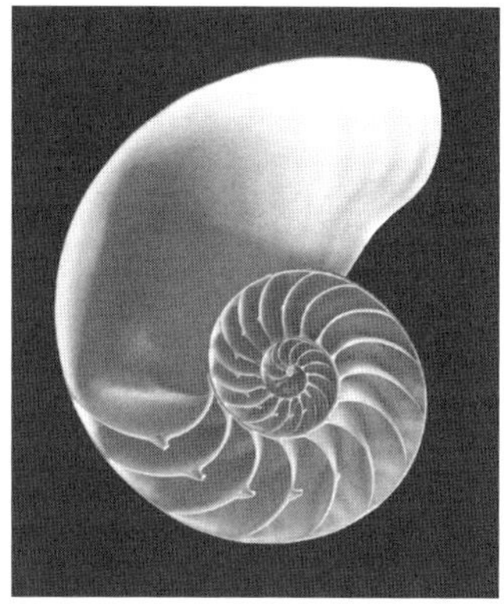

C3-10 Shell with spiral pattern
(colour plate x)

C3-11 Sunflower seed patterns
(colour plate x)

the other. Entire books have been written on this fascinating topic for children (Campbell, 2010) and adults alike (Posamentier & Lehmann, 2007).

Spatial spirals occur in space, such as the tendrils of a sweet pea as they climb in spirals up a string. Leonardo da Vinci recognized that leaves often arrange themselves in spirals in order to better share raindrops and sunlight (i.e., phyllotaxis).

C3-12 Spiral plant pattern (colour plate xi)

LEARNING EXPERIENCE
Exploring and Creating Spiral Patterns

- Encourage students to observe and make visual notations of natural forms that grow in spirals, such as the leaves on a stem or the centre florets of a sunflower or daisy.
- Students can create spirals by cutting a circle continuously and concentrically. Challenge students to use their spirals in a personal art image. How can spirals be incorporated into two- or three-dimensional artworks? For example, spirals can be used as a supporting structure for a mobile or sculpture, or as part of a relief collage.
- Count the spirals of the seed heads of the echinacea, daisy, or sunflower. How many rows go clockwise? How many go counter-clockwise? Draw or paint flowers that have seed heads that grow in spirals. Look at drawings and paintings done by artists such as Van Gogh and Georgia O'Keefe

(e.g., sunflower paintings). Students can start by drawing the spirals of seed heads, then the seed pods followed by the petals. Observe and draw one petal at a time until the flower is complete.

- Clay pots can be made using coils in a spiral formation. Students should consider the shape of the pot they wish to create. Study pottery created in places such as Africa, the southwestern United States, and India. Flowers and sunflowers can be used as a theme to create clay relief sculptures.
- Weavings can be done on a circular loom, such as those used to create the North American Aboriginal "dream catcher" and the South American "God's eye." Students can study various cultural weavings done on circular looms and create their own.

C3-13 Clay coil pots

Fractals

> *Why is geometry often described as "cold" and "dry"? One reason lies in its inability to describe the shape of a cloud, a mountain, a coastline, or a tree. Clouds are not spheres, mountains are not cones, coastlines are not circles, and bark is not smooth, nor does lightning travel in a straight line.* (Mandelbrot, *The Fractal Geometry of Nature* [1977], p. xiii)

Benoit Mandelbrot (1983) coined the term fractal for unique geometric figures that appear as fragmented spatial patterns, yet actually behave in organized, predictable ways. Found often in nature (e.g., coastlines, human lungs, certain plants), fractals are "self-similar," which means that they can be split into parts, each of which is (at least approximately) a reduced-size copy of the whole. The

C3-14 Fern fractal patterns (colour plate xi)

word comes from the Latin word *fractus*, meaning "to break." Fractals are the fundamental characteristic behind the growth of all complex organisms (Lesmoir-Gordon, Road & Edney, 2000). Romanesco is an edible cross between broccoli and cauliflower, and provides a visually striking example of how fractal patterns are formed in nature. A fern is another example.

LEARNING EXPERIENCE
Creating Fractals

◙ Simple fractals can be created with a pencil and paper, or on a graphing calculator. Follow the steps shown below to create your own Koch snowflake fractal pattern by simply inverting equilateral triangles of reducing sizes. Then observe and draw natural forms and plants (e.g., ferns) to identify whether or not the spatial patterns they exhibit can be classified as fractal (i.e., do smaller shapes/patterns in the plant echo the larger shapes/patterns?).

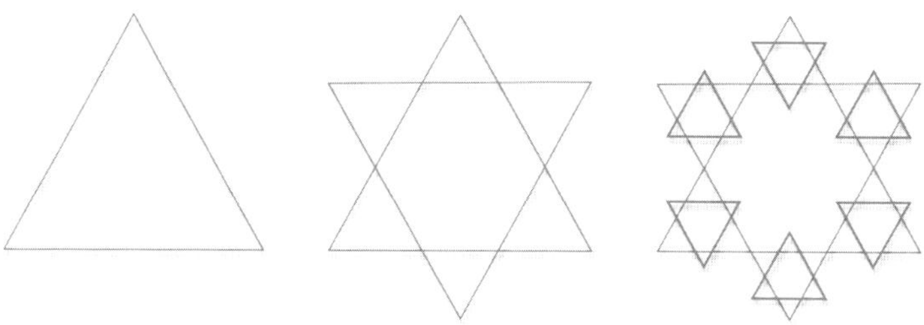

C3-15 Constructing a simple Koch snowflake fractal with equilateral triangles

LEARNING EXPERIENCE
Exploring Positive and Negative Shapes with Plants

Dr. Barbara Law of Nipissing University had her students create games for young children to encourage sensory-perceptual skills in physical education. One such game—Shape It—created by Meagan Bothan and Carley Foisy, reinforces the concept of figure-ground perception, or the ability of children to visually discriminate the main object that they are looking at (the figure) and everything else that forms the background (or ground), a skill much needed for human movement through space. We are pleased to highlight it here because not only does this game achieve the goals of the original health/physical education assignment, but it also can be an excellent way to have students practise positive/negative space recognition, which is very much a part of developing realism drawing skills (see Chapter 5).

▣ Have students present each other with shapes/images of flowers or plants that have to be duplicated by a partner using geometric paper shapes (either pre-cut, or cut out as needed) as negative space shapes (i.e., forming the "ground" or background of the image). In other words, the image is not reproduced by arranging interior pieces of paper to create the shape itself, but rather, exterior pieces of paper are arranged to reproduce an accurate negative shape/outline. While any kind of image could be used for this exercise, the complexity found in many plant outlines make these particularly useful objects for this activity.

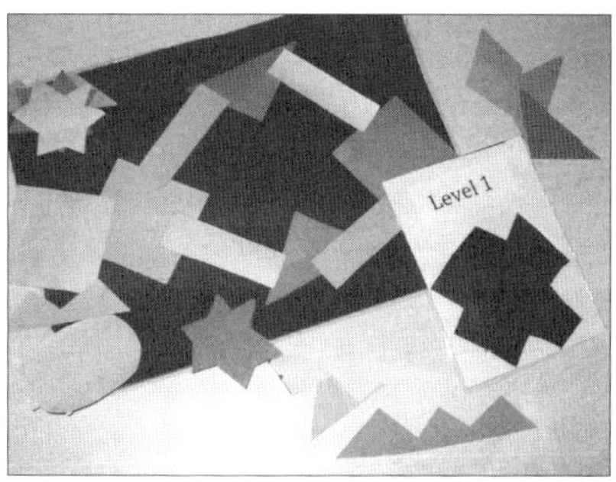

C3-16 Shape It, a game of visual perception (colour plate xi)

Flower as Power Image

Originally all flowers were wild. People developed and cultivated valued qualities of flowers for their appearance, shape, fragrance, and colour, and for their usefulness as dyes and for pleasant additions to foods. Flowers have also been used as medicine (e.g., as pain suppressants) and for their hallucinative abilities.

LEARNING EXPERIENCE

Observing, Drawing, and Reinterpreting the Flower

- Both art and mathematics require students to observe, recognize, categorize, and compare things visually and kinesthetically (Naested, 2010, p. 233). Have students draw a flower as accurately as possible. Suggest that students pretend they are a small bug crawling along the edge of the flower, the petals, and the leaves. Follow the edge into the flower or plant. As the eye moves along, the pencil should move on the paper at the same slow crawling rate. Count and record the petals and the spirals that are observed.
- Create a series of contour drawings of the flower as viewed from different angles (side, top, bottom), from close up, and an arm's length away.
- Draw a continuous line drawing of the flower.
- Draw the flower using only geometric shapes such as triangles, diamonds, or rhombuses.
- Draw the flower using only straight lines, curved lines, circles, and dots.
- Draw the flower using thick and thin lines. Attempt to use the thicker lines on the forms that are closest and thinner lines on forms that are farther away.
- Create a picture of the flower using colours as realistically as possible using paint or pastels.
- Flowers, petals, and leaves have curves. Some are open and some are

closed, some are concave and some are convex. Verbally describe and visually document your flower and leaves.

- Gently take a flower apart with a knife. Glue each part on stiff paper or cardboard and label them: anther, ovary, ovule, petals, pistil, pollen grains, pollen tube, sepals, stamen, stigma, and style. Redraw these parts as accurately as possible. Redraw these parts again using only geometric shapes.
- In a series of drawings, transform one of the flower or leaf drawings into something else (i.e., create a metamorphosis drawing). Each image should be a gradual transformation; for example, the drawing could show a flower with leaves transforming into a cup and saucer.

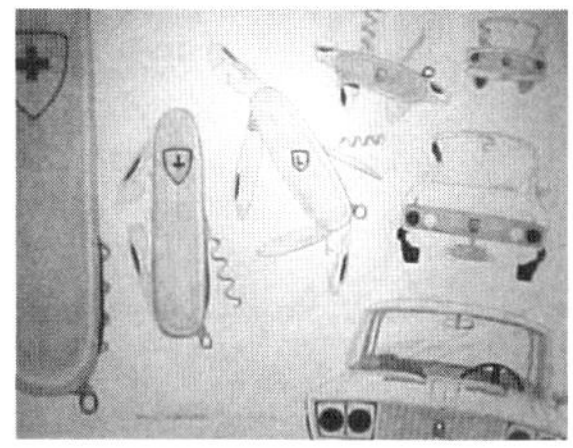

C3-17 Metamorphosis drawing

LEARNING EXPERIENCE

Create Macro Drawings with a Viewfinder

- With the use of a viewfinder, choose an area of the flower to enlarge. The viewfinder focuses the artist's attention on a portion of the flower. Draw the chosen area in great detail. This enlarged detail can be drawn with conté, chalk, or ink, creating a black, white, and grey image (i.e., three basic tonalities). Create a positive and negative (black and white) image of your drawing. This can be done with ink, paint, or by cutting the image out of coloured paper.
- Construct a viewfinder for use in determining what you wish to draw by cutting a rectangle out of a piece of paper or cardboard to create a "window." You might also use two "L-shaped" pieces of mat board, which can be adjusted to form a desired size of rectangular viewing window (as shown on next page).

C3-18 Beebalm flower detail for macro drawing (colour plate xi)

C3-19 Using a viewfinder made of two L-shaped mat board pieces (colour plate xii)

C3-20 Using your hands as a portable viewfinder (colour plate xii)

- Hands can also be used as a portable and effective viewfinder: twist the right hand and connect the index fingers with thumbs of the opposite hand as shown.

LEARNING EXPERIENCE

Flower Symbolism and Cultural Significance

- Flowers, emotions, and symbolism: Flowers are used to express feelings of love and sorrow, and to celebrate special events such as weddings, funerals, anniversaries, Valentine's Day, and birthdays. Certain flowers have been given symbolic meanings, such as the rose for love (Valentine's Day), the mistletoe for peace and love, the white Easter lily for purity (Christian), and the lotus as sacred (Buddhists and Hindus). Investigate other flowers and how they have been used in different cultures as symbols, and as inspiration for elaboration and decoration.
- Motif and logo: A motif is a dominant, often stylized or simplified symbol. A logo is an identifying symbol usually for advertising a company or organization. Flowers and plants have been used in motifs, logos, and symbols, where they have been simplified and stylized, decorated and elaborated upon, distorted and transformed. These images have been

used as symbols or decoration on such items as flags, banners, functional utensils, weapons, and clothing, as well as the human body. Have students search out company logos that have a flower or plant motif. From one of the logos, create a motif or logo for a wallpaper design, border, or stationery. Alternatively, students could create a flower design for a personal or class logo, flag, or banner in coloured paper, paint, or quilted or dyed cloth.

C3-21 Community quilt with blocks depicting local cultural scenes and symbols (colour plate xii)

C3-22 Stylized motif of flower in cast iron gate design

C3-23 Flower logo design work

The Grid

Dilatation (or dilation) is a mathematical term that means the scaling transformation of a geometrical figure, shape, or image. The new form has a similar shape to the original and may be larger or smaller (Fyfield & Blane, 1995). For centuries, artists have used grids, a network of horizontal and vertical lines, to transfer an image to a smaller or larger version of the original. A grid can also be used to estimate the area of a form, such as that of a leaf.

Distortion can be achieved through the use of a distortion grid. The horizontal and vertical lines on a distortion grid are not parallel or evenly spaced. They can be regularly or irregularly stretched, patterned, or curved.

LEARNING EXPERIENCE
Drawing and Distorting Plants Using Grids

- Students can collect various leaves and redraw them on a grid to estimate the area of the leaf. Using a distortion grid, students can enlarge or distort a flower or leaf.

C3-24 Grid drawing of a flower

LEARNING EXPERIENCE
Ferocious Flowers

- The theme of plants and flowers can become a springboard to creative the invention of mutants such as "Ferocious Flowers," as in carnivorous killer plants with barbs and sticky tendrils. Transform a flower in a drawing, painting, or sculpture (clay, papier mâché, playdough) into a "killer" plant for a hypothetical new film, a story or fairy tale, a comic strip, or a video game. Create patterns through exaggeration, distortion, enlargement, translation, overlapping, and repetition. Stories, poems, fairy tales and myths can be read to students beforehand, or students can develop their own fairy tale involving a Ferocious Flower as a writing project.

Symmetry

Interlocking shapes displayed in majolica tiles, inlaid wood, brickwork, carved stucco, stone pavement, sewn patchwork or printed fabric hold a special fascination for many people that goes far beyond the aesthetic pleasure that these patterns provide. (Schattschneider, 1993, p. 157)

Symmetry is the balance of a form in size, proportion, and position of parts. It means well-proportioned, or well-balanced. It is a mathematical concept as well as an aesthetic one (Stewart, 1995). Balance is a principle of visual composition. Proportion and balance appeal to both the artist and the mathematician. The human eye appreciates visual balance and pattern, which are often associated with beauty. Symmetry is found repeatedly in nature and is not only an aesthetic attribute, but also a practical one. For example, symmetry helps keep some plants from falling over to one side. Artists have used various forms of symmetry in their art images including bilateral (mirror image) reflection, glide reflection, and radial symmetry.

M. C. Escher (1898–1972), a Dutch mathematician, skilled graphic craftsman, and artist, graphically captured fantastic visual effects in his images of buildings and landscapes, and in his repeated patterns and tessellations. In 1921, Escher used a flower in his woodcuts, illustrating his interest in duality, or linear double contours, where a single line serves two visual purposes in a tessellating motif (e.g., the edge of a horse's mane on one side, but the edge of a horse's belly on the other). One of Escher's geometric prints, called *Beautiful*, was composed of alternating black and white diamonds with a central symmetry. When turned 180°, the print looks the same (Schattschneider, 1990a, 1990b).

C3-25 Ink drawing with bilateral, or mirrored, symmetry

LEARNING EXPERIENCE
Exploring Bilateral Symmetry with Plant Shapes

- After detailed study and the making of several drawings of flowers, fruits, vegetables, or trees, develop a simplified bold-shaped image of an object on a half sheet of dark (or light) paper. Cut out the shape on its contour line as well as any interior shapes. Save all of the leftover pieces of paper. Assemble the cut-out shape and any interior shapes on one side of a piece of light (or dark) paper. These are your positive shapes. On the other half of the paper, arrange all the leftover pieces; these are your negative shapes.
- Bilateral symmetry means that if you reflect the left half of an image, then you obtain the right half, or its reflection. Using the image developed in the last Learning Experience, create a mirror or reflected image. The mirror image can be created using dark and light colours and paints or paper to create strong contrast between the original and reflected image.

C3-26 Bilateral symmetry design

Tiling

Tiling (or tessellating) is the repeated use of a shape, or multiple shapes, to completely cover a two-dimensional surface without overlapping or leaving spaces or gaps between contiguous (adjoining) shapes. Islamic artists often used repeated colourful geometric shapes based on plants, in conjunction with their stylistic lettering called calligraphy.

C3-27 Naturally occurring geometric tiling pattern in honeycomb (colour plate xii)

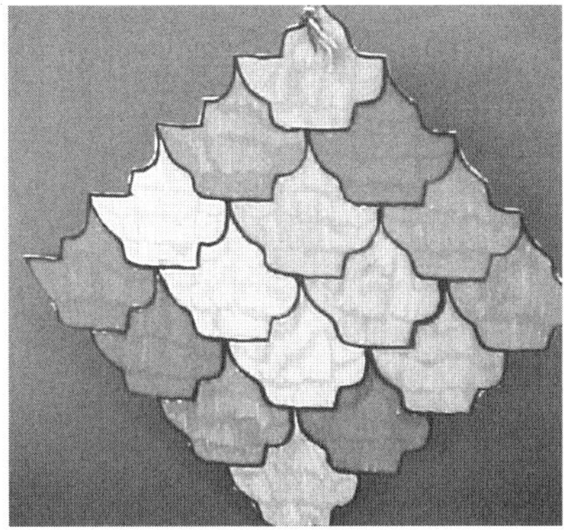

C3-28 Leaf tessellation pattern (colour plate xiii)

LEARNING EXPERIENCE
Exploring Geometric Shapes in Plants and Flowers

- After observing and drawing flowers and other plants, students can alter the main characteristic parts of plants into geometric shapes. These geometric shapes representing the plant can be repeated on paper, clay, or fabric to create a work of art.

LEARNING EXPERIENCE
Kaleidoscopic Designs

Sir David Brewster patented the kaleidoscope, which is a cylinder with mirrors that meet at an angle and run through the length of a tube. As the kaleidoscope is rotated, coloured-glass chips fall into various patterns, which are reflected on the mirrors. Students can create two-dimensional kaleidoscopic designs. Much like what they learn from cutting a round pizza or pie, students will gain understanding of fractions and angles, rotation and reflection.

A section of a design or drawing can be used to create a kaleidoscope effect when repeated in a circular fashion. After students have drawn and painted flowers, have them chose a "wedge" or portion of the drawing to be used as a "slice" for the pie shape. For the most successful results, choose shapes that touch the edges of the wedge to avoid too much empty space or a floating effect. Have students divide a circle into wedges. For eight wedges, use a protractor to mark points at 45° angle segments all around the circle (i.e., 360/8 = 45); for six wedges mark points at each 60° measurement (360/6 = 60). Students photocopy their wedge an adequate number of times to fill their circle. Wedges are cut out, placed in sequence, and decorated using various media.

C3-29 Kaleidoscopic flower designs (see plate xiii)

Trees

A tree is a living thing that grows from a seed, and the mass of foliage seen above the ground has a vast network of roots underground, equal in size. The trunk breaks out into branches and twigs in a general tapering of forms.

Throughout history, humans have used trees to feed, clothe, and shelter themselves. Trees are valued, revered, and sometimes even deified throughout parts of the world for their various properties of size, strength, and stability.

Many trees are among the largest plants in the world. The Douglas fir is the largest tree in Canada, some living for more than 1,300 years.

To draw trees well, artists should learn their patterns of growth, foliage, bark patterns, etc. Franck writes that drawing is a discipline by which one constantly rediscovers the world (1993). Every drawing teaches the artist something and can provide information or inspiration for visual expression in various art media. The following are learning experiences based on the study of trees. These can be added to, or changed, to fit studies of other plants, such as flowers.

LEARNING EXPERIENCE
Sketchbook Assignments Involving Trees

- Trees come in many different shapes, forms, and sizes. Study many trees in the local neighbourhood or surrounding countryside and take pictures of them. What types of trees are there? What are the names of the trees? Choose a tree and observe its size, shape, and colour.
- Trees are made up of different parts, such as branches, roots, trunk, leaves, or needles. Before attempting to draw a tree in full foliage, draw a single leaf, then a cluster, then a branch, and then a close-up of the bottom branches.
- Make rubbings of a tree trunk. Notice cracks and knots. How do these marks describe how the tree has grown? Can the texture on the tree trunk be described in mathematical terms (e.g., geometric forms, angles)? Compare the texture and colour of the bark of several trees.
- Observing how trees branch is a fascinating study. Look at the skeleton of a tree: its trunk and branches. Do

C3-30 Fall tree colours in North Bay, Ontario, Canada (colour plate xiii)

the branches grow up or down, or fan out? Look at the shape of your tree. Is the shape formed by the way the branches grow? Does the tree grow diagonally, slanted, straight, or curved? Naturally, branching patterns are somewhat random due to the influences of elements such as sun, shade, water, breakage, and collisions between branches (Ball, 1999).

- Draw the negative shapes, that is, the spaces between the branches, as geometric shapes (Harim & Love, 1998). Leonardo da Vinci suspected that there are algorithmic rules governing tree growth (Ball, 1999). In the 1890s, Wilhelm Roux tried to define the rules by identifying the following principles: "1. When the central stem forks into two branches with equal width, they both make the same angle with the original stem. 2. If one branch of the fork is thinner than the other, then the thinner branch diverges at a larger angle than the thicker. 3. Side branches small enough that they do not deflect the main stem appreciably diverge at angles between 70° and 90°" (Ball, 1999, p. 128).
- Benoit Mandelbrot coined the term fractals and believed that they are "the geometry of nature." Not all branching patterns are fractals, but "the fractal dimensions can be a useful measure for classifying self-similar

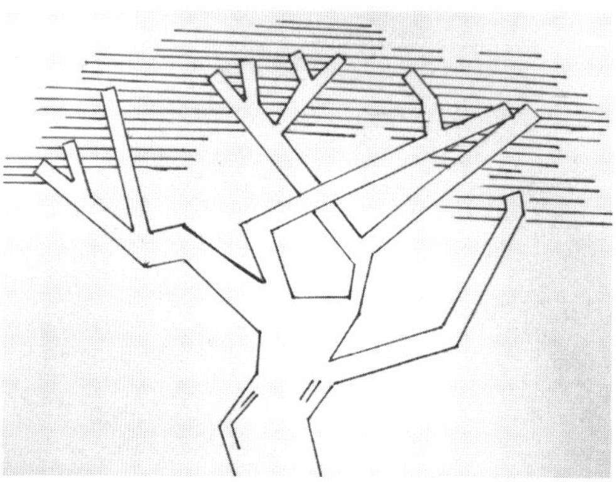

C3-31 Simplified tree drawing

C3-32 Tree painting (colour plate xiii)

structures" (Ball, 1999, p. 117). Observe a tree to determine whether it displays fractal patterns in its branching.

- Draw a silhouette of a tree: the basic outline without detail. The silhouette of the tree will be very different in various seasons and at different times of day as the sun shines through the foliage and branches (or gives only a muted, cloud-covered light). The silhouette will also change over the years due to growth and different environmental factors, such as heavy snowfall, lightning, or other accidents. Develop a simplified image of the tree. Draw it again on dark (or light) paper and create positive and negative spaces.
- Every year trees grow a little taller and spread a little wider. Roots grow deeper into the ground and out towards the drip line. The trunk gets wider and the branches get longer, with smaller branches coming out from them. Simplify the tree by drawing it with only a few simple straight lines.
- Draw a tree as realistically as you can: include the trunk, branches, and leaves. Use graphite pencil or ink.
- The texture of tree branches and leaves have aesthetic qualities, but also have functional properties. Donaldson (2003) writes that twisted growth might be influenced by the earth's rotation, and that the spiral textures help to transport water and other nutrients around the branch or tree. The texture one sees and feels on tree trunks and leaves may not immediately be believed to be a mathematical consideration. Observe and draw the texture of various parts of a tree, such as growth rings and spiral bark, branches, and leaves. Describe the tree (e.g., texture, branches, growth rings) in mathematical terms (e.g., number, size, shapes).

C3-33 Tree leaves watercolour painting (colour plate xiv)

LEARNING EXPERIENCE
Measuring Tree Growth

Growth is a progressive development or increase, and measurement is the process of calculating one or more of an object's qualities to find a standard or to compare. Studying trees is a great experience for students to understand the concepts of growth and measurement and to record their findings. The following learning experiences also assist in focusing students' attention on the details of trees in order to broaden their understanding and enrich their visual representations.

- Students can measure the growth of a tree throughout the school year. Measure a student's height. Place a mark on the tree trunk at the student's eye level with outdoor paint. Measure the mark from ground level and record. Take the measurements again at the end of the school year (likely both the student and the tree have grown in height).
- Measure and record the circumference of the tree at the painted mark. Measure the tree again at the end of the year (along with the student's height). How many centimetres has the tree grown thicker? Calculate the percentage growth (Harim & Love, 1998). If possible, observe a cut section of a downed tree. Identify its parts: cambium, cork, heartwood, outer bark, phloem, sapwood and xylem. Make note of the rings and measure the distance between the rings. What can the rings' sizes and shapes tell you about the tree? The formal study of tree life through ring observation is known as *dendrochronology*.

C3-34 Dendrochronology (colour plate xiv)

- To measure the height of a tree, have one person stand against the trunk of a tree. Hold a stick vertically and mark the spot on the stick when you can view the person from toe to head. The height of the person becomes the standard length of measurement. Sight measure the tree. How many lengths high or units tall is the tree? Let's say the tree is three person measurements high. Next, accurately measure the person who stood beside the tree. Let's say the person is 150 cm tall. Multiply this height by the factor of three: 150 × 3 = 450 cm. The tree is therefore approximately 450 cm tall.

C3-35 Using a pencil to compare proportions in landscape drawing

- Another way of finding a tree's height is to use indirect measurement. On a sunny day, measure your own height and cast shadow, as well as the cast shadow of the tree. Then using similar triangles, we can set up an equation and solve for the unknown distance. So, for example, if you are 1.5 m tall and cast a shadow of 1.6 m, and a tree casts a shadow of 15.4 m, then we would write the proportion

$$\frac{1.5}{1.6} = \frac{x}{15.4}$$

which then is rearranged as

$$x = \frac{1.5\ (15.4)}{1.6} \approx 14.4\text{m}$$

Conan Doyle's famous Victorian detective character, Sherlock Holmes, used a similar method in the classic tale *The Musgrave Ritual* to help solve a perplexing case involving the height of an ancient oak tree and the distance to a buried treasure.

C3-36 Sherlock Holmes

- Shadows of trees change throughout the day depending on the angle of the sun. Shadows are longer in the morning, shorter as the sun is higher in the sky around noon, and by late afternoon, the shadows are longer again as the sun moves lower in the sky. On a sunny day, take students outdoors in the morning to draw a tree. Sight measure the shadow of the tree in the morning and draw the shadow. Go out again at mid-afternoon and again at the end of the day; each time, have students sight measure and draw the shadows. Discuss what students learned about shadows from this experience. How have the shadows changed? This is important information for artists when drawing and painting landscapes.

LEARNING EXPERIENCE
Observing and Creating Tree Designs

- Geometric format distortion: Distort a tree as Mondrian did in many of his paintings. Distort it to fit into a geometric format, such as a circle, rectangle, square, or oval.
- Create paintings of trees using various colour schemes, such as warm and cool colours, to create different emotional effects. Draw a tree in various seasons. Use tree patterns to create a motif, a border pattern, or mosaic tiles. Create a large painting or mural and incorporate trees into the mural.
- Three-dimensional sculpture: Using the tree as a theme, create a three-dimensional sculpture or model of a tree form using various sculptural media such as clay, papier mâché, wire, cardboard, or playdough. Have students measure or estimate the amount of sculptural material they will need for this project.
- Weaving: The branches of trees can be used to make a loom for weaving. The branches can be bent into a circular form or a more rectangular shape to make the frame. Wrap yarn around the branches, and then string the

yarn either across the circle or up and down the rectangular frame. These "strings" become the warp of the loom. Weave threads, yarn, and other material across (in and out, over and under). These threads become the weft (or woof). The coloured threads can be woven into patterns by counting the warp and weft threads, such as by weaving two warp threads over and two under. For further information on weaving, please refer to Naested (2010).

C3-37 Weavings like a spider's web can be done with a circular frame

Leaves

> *Eddie often took oak leaves and traced around them. He loved the shape—there were so many more zig-zags than in the leaves of the plane trees, or the poplar. He liked the chestnut leaves too, and he never played the silly game that the others did at school—peeling away the green bits to see who could have the most perfect fillet, like a fish with no flesh, only bones. Eddie liked the texture of the leaves.* (Binchy, 1992, p. 115)

The silly game that Eddie's schoolmates played by peeling away the green bits of the leaf revealed the delicate system of veins (that fascinated them). We generally consider leaves as being produced by deciduous trees such as the oak, maple, birch, and chestnut. The shape of their leaves can be broad or lobed, or divided like fingers into leaflets. It turns green due to chlorophyll in the spring, and yellow in the fall due to carotene. Some turn red due to the excess of sugar. "The outside of a leaf is waxy, to prevent the evaporation of water, and the underside is matte and spongy to absorb carbon dioxide and exhale moisture" (Taylor, 1974, p. 12).

According to Ball (1999), the arrangement of leaves on a plant stem follows a system of special mathematical forms. Leaves commonly group up the stem in a spiral formation, each leaf displacing the one below by a fairly constant angle of offset close to 137.5°. Another arrangement called distichous places leaves on opposite sides of the stem, where the leaves often wrap around the stem with an offset angle of 180°. The third pattern Ball describes is the whorled pattern, which has two or more leaves at regular intervals up the stem. Here the leaves are often 180° apart, offset at angles of 90° from the two leaves below (Ball, 1999, pp. 104–5). The arrangement of leaves helps each leaf receive sunlight and rain.

LEARNING EXPERIENCE
Exploring Leaves

- Examine various plants and trees to determine their leaf arrangement: spiral, distichous, or whorled.
- Conduct a leaf vein study. You needn't peel away the flesh of the leaf in order to get to know the veins of the leaf. Pay close attention to detail and draw the leaf, one vein at a time, perhaps using a magnifying glass. Draw precisely what you see, varying the thickness of the lines being drawn to reflect the thickness of the veins.
- Feel the texture of a leaf of your tree and take some rubbings of it. Place a leaf under paper and rub over top with a crayon or soft pencil. Place many leaves in many positions to create a composition. What are some ways to create emphasis and movement in the composition? When the rubbing has been completed, a light paint wash can be brushed over the crayon or wax rubbing. The paint is resisted where the marks from wax or crayons have been made. If soft pencil has been used, the light wash will allow the pencil marks to show through. You might also use white or coloured crayon or pastel and overlap the rubbings. Brush a light wash

of darker coloured paint over top to create a "batik" effect.

- Take a rubbing, brush on a light paint wash, let the paint dry, and then take another rubbing and bush on another paint wash. This can be repeated many times to give the picture depth through overlapping and the buildup of colour from the crayons and paint.

C3-38 Maple leaf vein patterns (colour plate xiv)

- Collect leaves of many trees, especially in the spring and fall. Draw a leaf in detail. Observe how the main rib tapers from the base, while the other ribs emerge from the centre rib in a regular tapering pattern. What leaves are octagonal in shape? Are there any other shapes? Note the variations of colour. Press the leaves for future reference.
- Make an exact drawing or outline of a leaf on graph paper. Record the coordinates on grid paper and then "connect the dots."
- Trace a leaf on one centimetre grid paper. Find the area of the leaf by using the coordinates. Find the number of whole squares and the number of fractional squares. Colour in the areas. Add the whole and fractional squares (by adding the fractions) and approximate the total area of the leaf in square centimetres.
- Printmaking: Roll ink or paint on a leaf and create a repeated pattern, translation, five-fold rhythm, pentagonal grouping, or radial design. Using cardboard and string, create a relief print of the leaf. Observe and draw tree roots (i.e., pulled

C3-39 String relief print of trees

from the ground, or by way of photograph) and make a string relief plate of the drawing. Make a print of the roots using ink.

- If access/permission is possible, tap a maple tree and collect sap in a bucket during the spring runoff. Boil the sap to evaporate much of the water. Record how much sap is required to produce maple syrup: did your ratio approximate 40:1?
- Needles and pine cones: The needle-like "leaves" that are not shed in winter are produced by coniferous trees. The most common tree group, the conifers, includes the pine, hemlock, cypress, redwood, juniper, yew, fir, and spruce, which all have needles and pine cones. Pine cones have bracts, or scales, which are ovule- or pollen-bearing and vary in shape from oval to more elongated forms. The scales grow in patterns that spiral around the cone, often in a Fibonacci sequence (in Figure C3-40, note the 13 spirals that emanate from the centre of the cone). Collect as many variations as possible, observe and draw their shapes and scales, and count the number of spirals in the growth pattern.
- Cactus needles: Cactus needles are also considered leaves. Observe and draw the pattern that the needles of the cactus grow from the fleshy cactus trunk, stems, or arms. What geometric patterns do they form?
- Seed pods: Flowering plants produce seeds, nuts, and fruit. The flowers are a great source of inspiration for artists and mathematicians, not to mention biologists. Fruits often contain an outer shell or enclosure for the seed, or seeds, that are arranged in a particular way. The "seed container" can be spherical, elongated, or ridged (Taylor, 1974). Collect various seed containers. Draw their outward appearance.

C3-40 Pine cone scale patterning (colour plate xiv)

Cut the container in order to expose the arrangement of seeds. Draw this section, carefully capturing the shapes and textures using various soft and hard pencils (6B, 4B, HB, 2B, 2H, 6H). Create a logo or motif based on the seed pod.

Landscapes

Three of the most common subjects for artists throughout art history have been still life (e.g., fruit, clothing, objects), portraits (human and animal models), and landscapes. The careful observation of landscapes provides the artist with an opportunity to truly appreciate the beauty of natural forms, and the way in which natural light and other environmental conditions can change the landscape's appearance. Impressionist artists such as Monet and Pissarro would paint many versions of the same landscape in order to capture the varying affects of light, time of day, and atmospheric conditions in their "impressions" of the natural world.

LEARNING EXPERIENCE
Creating Landscapes

- Foreground, middle ground, and background: Repeat a drawing of a tree to create a forest by overlapping the trees. Draw the trees in a relationship with one another. Create the illusion of depth by placing the trees at different levels on the picture plane. The bottom of the picture plane becomes the foreground. The middle of the picture plane becomes the middle ground. Higher up on the picture plane, above the middle ground, becomes the background. Consider varying the size of the tree. The shorter trees will appear to be in the background.

C3-41 Landscape painting with foreground, middle ground, and background (colour plate xv)

C3-42 Landscape photograph of trees and lake (colour plate xv)

- The golden ratio: Another pleasing way to compose a two-dimensional image of a landscape is using the golden ratio, or golden rectangle. Have students measure the width (shortest side), multiply this number by 1.618, then mark and draw this length as the other dimension of the rectangle (i.e., length). They can then mark where the width would form a square along the length. The remaining distance from here to the end of the length becomes the next smaller square in the golden rectangle, and so forth and so on, until the "whirling square" pattern becomes very small in the eye of the golden spiral (see illustrations in Jarvis, 2002, 2007). When drawing, you can use these construction lines to purposefully place or cluster your images, or use the horizontal lines as foreground, middle ground, and background or to choose where to place the horizon line in the landscape.
- To analyze any existing rectangle shape to see if is it "golden" or not, simply divide the longer side by the shorter side to see if the resulting value is close to phi, or 1.618. For example, a sheet of foolscap paper, available in most schools, is very close to being "golden" when its length

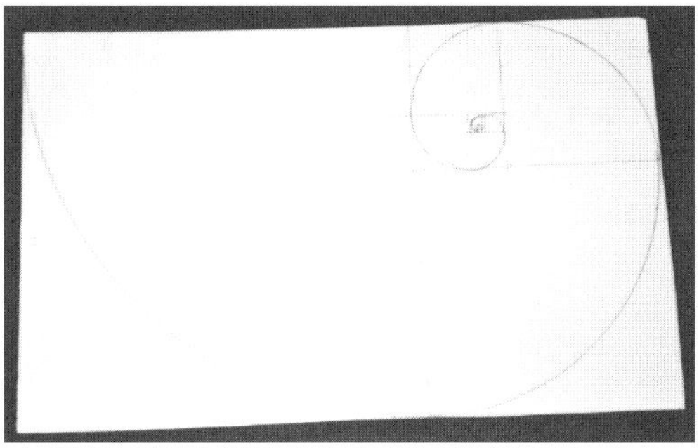

C3-43 The Golden Rectangle featuring "whirling" square pattern (colour plate xv)

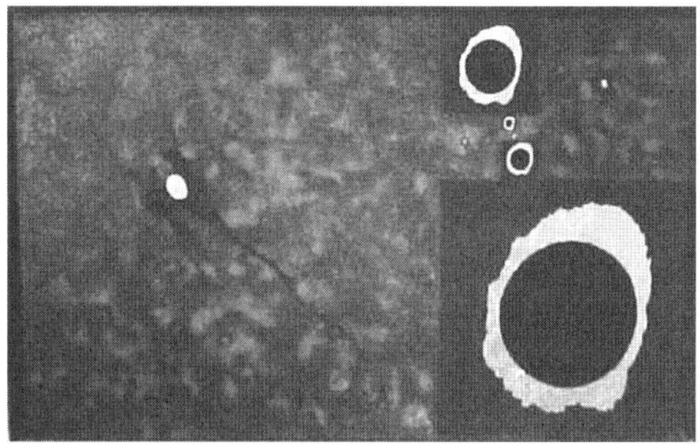

C3-44 Golden rectangle painting with iterating neural dendrites (rise) and solar eclipse (fall) images (colour plate xvi)

is divided by its width. Have students look at landscape paintings from artists throughout history. Did they use the golden ratio? Analyze several rectangular household objects to see if they are "golden."

Rainbows

> *As sunlight passes through raindrops, the density of the water causes the light to slow down ever so slightly. Different frequencies slow more than others, causing the various colours of the spectrum to separate and leave the drops at different angles, exposing the individual colours that make up white light. This is much like what happens to light that travels through a glass prism, except that, with raindrops, the light is also reflected back to the observer's eyes. The seven colours of the rainbow are, from outside to inside, red, orange, yellow, green, blue indigo, and violet. In a secondary rainbow where light is reflected twice, the colours are in reverse order. You may also notice an area between a primary and a secondary rainbow that is darker than the surrounding sky. This is known as Alexander's Dark Band and it is dark because light from this area is not reflected to the observer's eye.* (Dunphy, 2004, p. F8)

How often have you observed a rainbow in the spring, summer, or fall? Or a sundog on cold winter days? In order to paint a rainbow into a landscape, it is important that artists observe and gain some understanding of rainbows. In mathematics and science the rainbow is a prism; optically, the rainbow is a distorted image of the sun. Raindrops perform this distortion through reflection and refraction. Different colours are refracted by different amounts when light passes from one medium, such as air, into another, such as water. Rainbows are rarely seen at noon since the viewer needs to see raindrops sunlit at a 42° angle.

We often forget what colours are in the rainbow and the order of the colours. An easy way to remember (mnemonic device) is the silly name: "ROY G. BIV." The colours appear from the top of the rainbow to the bottom, or from outside to inside, as red, orange, yellow, green, blue, indigo, and violet.

It was Isaac Newton who, in 1666 (the year known as the "*annus mirabilis*" or "year of miracles/wonders"), during the school cancellation resulting from the Black Plague in England, discovered not one, but four major inventions/theories in the fields of calculus, motion, optics, and gravitation that would change the world for the next three centuries. One of these discoveries was light refraction, the colour spectrum that is seen in the rainbow, but which can be reproduced using a glass prism. The rainbow spectrum of colours can be viewed through the refracted light of prisms. Prisms are found in nature such as quartz crystals and waterfalls, and can also be designed by humans, like the glass prisms you can buy in a science curriculum shop or catalogue. Prisms can be named according to the shape of their bases. "A right regular prism is a solid that has a pair of congruent regular polygons for its bases and congruent rectangles for the rest of its faces" (Jacobs, 1970, p. 239).

The rainbow is an ancient, universal symbol. It often represents the connection between humans and their gods, or peace promising sunshine after a rain. The Australian aborigines call their goddess of creation the "Rainbow Serpent Mother." In Scandinavian mythology, the rainbow was a bridge between the gods and the earth. In the Bible, a rainbow created by the Hebrew God, El Shaddai ("God Almighty"), showed Noah and his family that the biblical flood was over, and that the earth would never again be destroyed by water. The colours of the rainbow are also associated with Hindu chakras of the human body and crystals: for example, red for the base or root, orange for the navel, yellow for the solar plexus, green for the heart, blue for the throat, indigo for the third eye, and violet for the crown. The colours also have symbolic meanings such as red for life, orange for healing, yellow for sun, green for nature, blue for sky or peace, indigo for harmony, and violet for spirit. The seven colours of the rainbow may have given the number seven magical or mystical strength in some cultures. Stories and myths surround the idea of the rainbow as a sign of goodness, luck, or wealth, as in, for example, the "pot of gold" at the end of the rainbow myth.

LEARNING EXPERIENCE
Exploring Rainbows

- Students can explore rainbows directly by creating conditions that allow for their observation. Fill a shallow dish with water. Place a mirror in the dish so that it rests against the side of the dish. Secure the mirror in place with a weight. Shine a flashlight on the part of the mirror that is below the surface of the water. Hold a piece of stiff white paper above the flashlight. Move the paper to see a rainbow. Students can also observe rainbows as they appear in nature, in the sky or through sprays of water. Through these activities, students will gain a greater understanding of how colours reflect and refract light, and may be more thoughtful when including rainbows in their landscape pictures. Using watercolours, mix colours and develop landscape or fantasy paintings that involve rainbows (featuring either the ROYGBIV colour spectrum order found in nature, or an imagined colour order of their own design).

LEARNING EXPERIENCE
Rainbows, Culture, and Illustrated Creative Writing

- Depending on the grade level and the social studies unit, research a culture and its art images and stories pertaining to rainbows. Create or develop your own stories, myths, or poems relating to the culture under study. Develop visual images to illustrate your written work, considering elements of art/design (colour being one), along with shape, tone, texture, and line. Develop the visual statement by also incorporating one or two principles of composition/design.

Clouds

Mandelbrot coined the term *fractal* for unique, beautiful geometric figures, found often in nature, that appear fragmented, yet actually behave according to predictable and self-repeating mathematical patterns. Fractals are generated by an iterative or repeated process. When one looks at a part of a cloud, the part often looks much the same as the whole. Cloud figures can sometimes be described using fractal mathematics. Clouds are named due to their shape, altitude and precipitation. Clouds can be cumulus or puffy clouds; stratus or gray-layered clouds, or cirrus wispy thin clouds high in the sky. Thunderhead or thunderstorm clouds are cumulus clouds. Clouds can be found high in the sky, 6,000+ m above ground, in the middle zone between 2,000 and 6,000 m, or low, below the 2,000 m mark.

LEARNING EXPERIENCE
Observing and Depicting Cloud Formations

- Observe and draw clouds. Attempt to name the clouds and estimate approximately how high they are in the sky. Discuss the spatial patterns they exhibit.

C3-45 Rainbow colours (colour plate xvi)

C3-46 Clouds and Great Lakes vista on Lake Huron near Goderich, Ontario, Canada (colour plate xvi)

- Phil Chadwick, an environmental weather expert, has applied his knowledge to the study of the artworks of Tom Thompson, J. E. H. MacDonald and other Group of Seven artists. Chadwick believes that it was the artists' "profound understanding of weather that helped them be in the right place at the right time to catch storms, cloud formations or jet streams" (Fischer, 2006). Students can observe landscape painting from the Group of Seven with specific attention to cloud formations, the understanding the artists had of cloud formations, and the use of painting techniques to represent clouds. Students can observe cloud formations in photographs or outside "on location" and create drawing and paintings based on these observations.
- Many artists (e.g., Van Gogh) have used clouds in their landscape paintings for emotional effect. After studying clouds, visually illustrate these natural phenomena to add emotional impact to landscape paintings. Consider the elements of colour, texture, and line and the principles of composition such as movement and emphasis.

LEARNING EXPERIENCE
Investigating the Natural Environment in Literature

- Literature, poetry, art and mathematics: Read about trees in books such as Joyce Kilmer's *Trees* and Shel Silverstein's *The Giving Tree*. Individually or in groups, create a poem, story, myth, or fairy tale about a tree. Furthermore, investigate stories and myths that involve rainbows and clouds. Attempt to use mathematical and visual art terms as often as possible in the writing of your own works of prose or poet-tree!

Summary

Chapter 3 has looked at the incredible order of the natural world as seen in flowers, leaves, and trees, along with rainbows and clouds. The incredible order in the way these natural forms develop and how this order can be interpreted through mathematics, geometry, science, and the visual arts never ceases to amaze the careful observer of nature. Artists and mathematicians throughout history have been inspired by the naturally occurring forms, shapes, patterns, and colours in the environment. Many mathematical and artistic learning experiences were presented in this chapter. They required the learner to see and touch, and to really observe the natural forms in order to gain greater awareness and understanding. Many art techniques and media are mentioned in this and other chapters. While the appendices offer brief descriptions of key terms in mathematics and visual arts, further investigation (print, web) into the art techniques and media discussed in the book is encouraged.

4 Mathematics and Art in Fauna of the Land, Sea, and Skies

C4-1 Tiger head symmetry, positive and negative balance (colour plate xvii)

Tyger! Tyger! burning bright in the forests of the night,
What immortal hand or eye could frame thy fearful symmetry?
In what distant deeps or skies burnt the fire of thine eyes?
On what wings dare he aspire? What the hand dare seize the fire?
And what shoulder, and what art could twist the sinews of thy heart?
And when thy heart began to beat, what dread hand, and what dread feet?
What the hammer? What the chain? In what furnace was thy brain?
What the anvil? What dread grasp dare its deadly terrors clasp?
When the stars threw down their spears, and watered heaven with their tears,
Did he smile his work to see? Did he who made the Lamb make thee?
Tyger! Tyger! burning bright in the forests of the night,
What immortal hand or eye dare frame thy fearful symmetry?

—William Blake, 1794

Introduction

It is with wide-eyed wonder that humans throughout recorded history have looked upon the planet's fauna and have been amazed by the varieties of land-dwelling animals, aquatic life forms, insects, and birds of flight. Many of these creatures have served special roles in relation to human beings, sometimes functioning as companions and protectors, and at other times providing humans with valuable materials. In this chapter, readers will gain a greater understanding of the relationship and connection between mathematics and visual arts through the study of various fauna. Learning experiences for mathematics and visual arts, as well as key concepts and images for process and product, appear throughout the chapter.

C4-2 Butterflies (colour plate xvii)

Describing the Bengal Tiger

The stance had something of a pose to it, as if it were an intentional, even affected, display of mighty art. And what art, what might. His presence was overwhelming, yet equally evident was the lithesome grace of it. He was incredibly muscular, yet his haunches were thin and his glossy coat hung loosely on his frame. His body, bright brownish orange streaked with black vertical stripes, was incomparably beautiful, matched with a tailor's eye for harmony by his pure white chest and underside and the black rings of his long tail. His head was large and round, displaying formidable sideburns, a stylish goatee and some of

> *the finest whiskers of the cat world, thick, long and white. Atop the head were small expressive ears shaped like perfect arches. His carrot face had a broad bridge and a pink nose, and it was made up with brazen flair.* (Martel , 2001, p. 167)

This quote from Martel's (2001) *Life of Pi* illustrates the writer's rich vocabulary and keen eye for the beautiful detail found within the animal he is observing. He describes the tiger's colour and pattern, but also the shapes of rings, circles, and arches, as an observant artist and scientist.

Humans have always looked at animals of the land and sea with wonder and have written poems, stories, fantasies, myths, and songs about them. Creatures have functioned as companions, listeners, and human protectors. Their feathers, fur, and skin have been used by humans to keep warm and to decorate human bodies. They have also been used as food.

Artists have looked to animals for inspiration, for visual expression on clothing, weapons, shelter, masks, totems, sculptural forms, and kites, for personal and group identity, and as forms of deity to be worshipped. Animals have also been used to express feelings of power, love, and relationships and to celebrate special events such as birthdays, or on family crests, Asian calendars, and totem poles.

Drawing Animals

Animals are of many types, shapes, and sizes. They walk, crawl, run, swim, and fly. There are wild and tame animals, land and water animals, and warm-blooded and cold-blooded animals. Scientists have classified animals into phylum, then into classes, orders, families, genera, and species, depending on their similarities and differences. Throughout history, artists have worked closely with scientists, or functioned as artist-scientists themselves, as they documented the natural world. These artists have visually captured animal structure and skeleton, body covering and colouration, habitat, adaptations, natural enemies, and their environment, food sources, and young. Your students can become avid and experienced artist-scientists.

LEARNING EXPERIENCE
Sketchbook Assignments

The following are sketchbook assignments that can be revised, expanded, or made more specific according to your students' studies and/or interests. Materials needed may include a sketchbook or paper attached to a board or heavy cardboard; drawing materials such as pencils, crayons, pastels, felt pens, paints, or pens and, possibly, a camera.

C4-3 Cat sketches

- Choose an animal to study. Ask questions, read about, and make notes about this animal's natural habitat and behaviours.
- Choose a picture of the full-frontal view of the animal's head. Cut the image in half and draw the other half. Measuring will be needed with either pencils or rulers. Use a variety of pencils (i.e., from soft 6B, 4B, and 2B to harder 2H, 4H, and 6H) in order to create the light and dark tones of the image and to create texture such as hair or fur.
- Make three to four drawings of detailed parts of the animal (e.g., head, eyes, feet, and ears).
- Sketch the animal in terms of the geometric shapes of the various body parts. Look for cubes, rectangles, spheres, circles, cylinders, parallelograms, and triangles.

- Draw several complete sketches of the animal in characteristic poses. Record the animal's colour using coloured pencils, pastels, crayons, or paint.
- Do several quick gesture drawings of the animal in order to capture the proportions. In gesture drawing, you do not draw what the animal looks like, but focus more on what it is doing. Overlap the drawings as the animal's movement is captured on paper, creating an illusion of movement. This can also be an exercise in size estimation, as students estimate the various sizes of parts of the animal.
- Create contour and simplified drawings of the animal in order to study proportions and measurement. Contour drawing means drawing the edge of an object, including the folds and indentations, as a way of coordinating the hand and the eye—what is seen with the eye and the movement of the drawing instrument. Contour drawings are line drawings that are developed form-by-form, detail-by-detail, to arrive at a completed form. As the eye follows the contours of the animal, the hand draws a line, trying to "feel" the edge of the animal visually. Suggest that your students focus on one point of the animal and place their pencil on the paper, which will correspond to the visual focal point on the animal. As students move their eyes slowly along the contour of the animal, they should move their pencil on the paper at the same time.

C4-4 Large snapping turtle (colour plate xvii)

C4-5 Turtle cardboard relief print

C4-6 Turtle Styrofoam etching print

C4-7 Bear portrait drawing

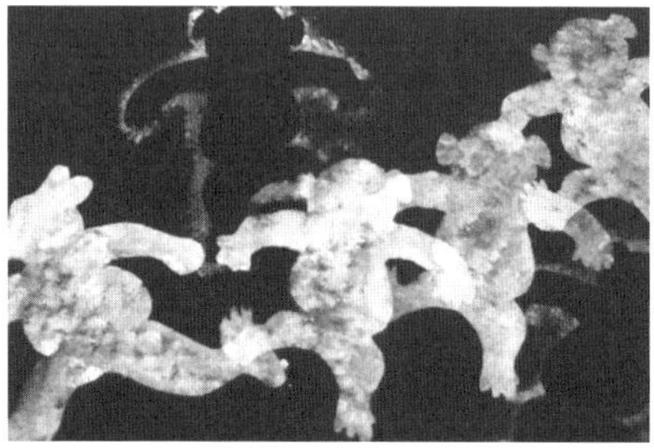

C4-8 Dancing bear stencil print, positive & negative

- Make note of the animal's sounds and the noises it produces. Can these sounds and noises be depicted on some form of a graph?
- Watch the animal for 5 to 10 minutes and make a map of its movement in its enclosure. Take pictures of this animal in different poses and from different angles.
- Look for articles about the animal in magazines, and photocopy the articles and images. Look for images in magazines, newspapers, and the Internet that remind you of this animal, its colouration, surface, coat, demeanour, etc.

C4-10 Elephant contour drawing

C4-9 Elephant riding in Chiang Mai, Thailand (colour plate xvii)

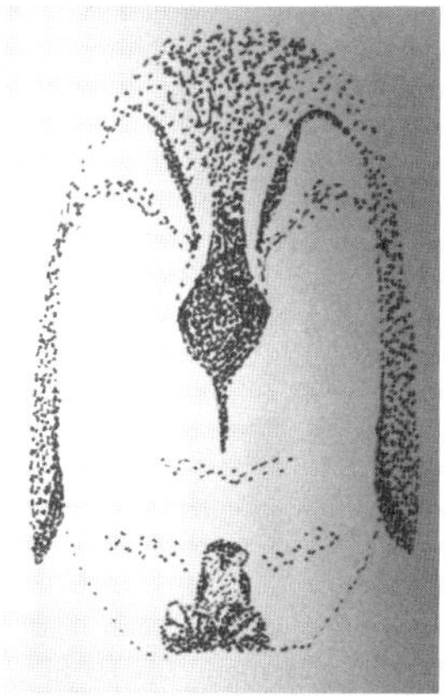

C4-11 Penguin drawing with pen and ink stipple technique

C4-12 Bird relief print

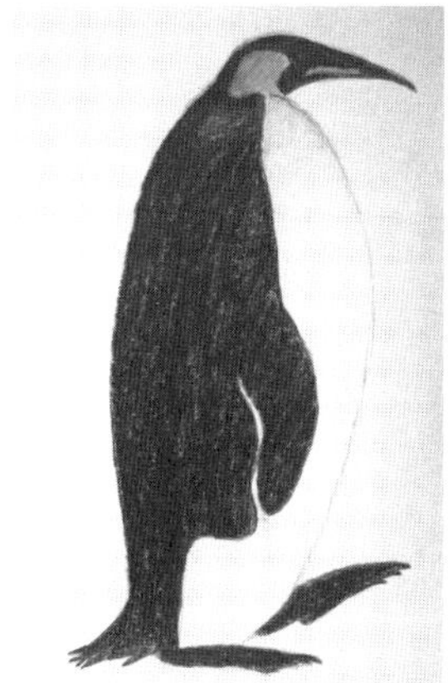

C4-13 Penguin colour drawing
(colour plate xviii)

- Hybrid animals are a combination of two different animals. While certain real examples exist in our world through crossbreeding (e.g., the liger is a hybrid cross between a male lion (*Panthera leo*) and a tigress (*Panthera tigris*)—it has parents with the same genus but different species), imaginary forms of hybrids have often been popular in film and other media. Have students use their drawing, painting, or sculpting ability to create and name an invented hybrid creature.
- Classify animals by drawing details of their feet, body covering, and colouration, as well as researching their habitat and food. These observations and related drawings can be used as information to create prints, paintings, murals, dioramas, or sculptural forms, or as a basis for works of the imagination that students can develop into personal statements.
- Transforming an animal portrait: After drawing an animal face from observation, students can take one of their drawings and gradually distort the features by stretching parts in several phases through transformations or metamorphosis to create an entirely different image (either a new creature or an inanimate object).

Wavy dabs of black circled the face in a pattern that was striking yet subtle, for it brought less attention to itself than it did to the one part of the face left untouched by it, the bridge, whose rufous luster shone nearly with a radiance. The patches of white above the eyes, on the cheeks and around the mouth came off as finishing touches worthy of a Kathakali dancer. (Martel, 2001, p. 198)

LEARNING EXPERIENCE

Animal Portraits Using Measurements and Comparison

Drawing from real life is a great experience, whether by going to the zoo or bringing live animals into the classroom. However, photographs and other pictures, as well as stuffed animals or animal toys/models, can also be used. Using a ruler or the end of a pencil, students can sight measure or calculate the size of the various animal parts and produce a portrait.

1. Measure the length of the animal's eye using a ruler or with end of a pencil (this eye becomes the standard "unit").
2. Draw the eye to the right or left of the page.
3. From the eye duct, measure one eye length horizontally in the other direction, and here begin to draw the second eye.
4. Measure the nose (the tip of the nose will be approximately one eye-length down from the bridge of the nose).
5. Measure and lightly draw a vertical line down from the tip of the nose to the lips (a half-eye length down).
6. Draw vertical lines down from the mid-point of each iris (this will be approximately where the corners of the mouth will end).
7. Note that the top of the ears are usually horizontal to the eye (or slightly above) and that the bottom are approximately horizontal to the bottom of the nose.

Grid Drawing

Albrecht Dürer (1471–1528) was an artist who used the grid to assist him in depicting his subjects. When asking your students to copy from a picture, photograph, or artist reproduction of drawings or paintings, have specific objectives in mind, such as emphasis on drawing and shading techniques or understanding the mathematical idea of using the grid (proportion, enlargement, magnification, distortion). Drawings can be enlarged (grid up) or reduced (grid down). The use of distortion grids can produce interesting, inventive results.

C4-14 Elephant drawing with rectangular grid

C4-15 Giraffe drawing with rectangular grid

LEARNING EXPERIENCE
Grid Drawings Using Dilatation (Scaling)

- Students select an animal to draw or trace on paper. A grid is then laid overtop of the drawing and an enlarged version of the grid is drawn on a larger sheet of paper or any large flat surface (e.g., wall, ceiling, billboard, vehicle door). Students may then enlarge the image from the smaller to the larger grid by carefully noting where and how the various lines of the drawing intersect within the grid squares.

C4-16 Bird

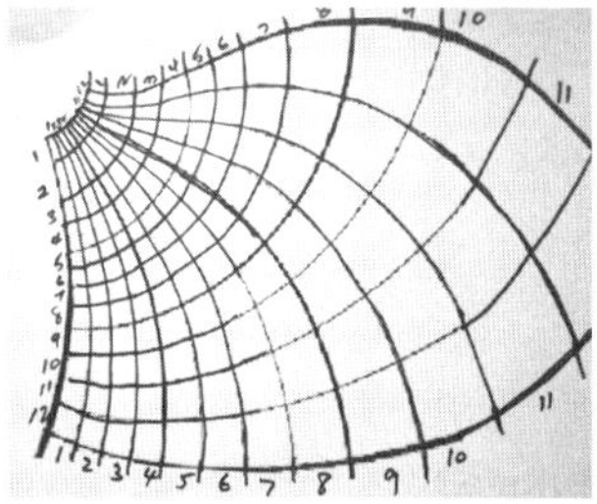

C4-17 Sample distortion grid for drawing

C4-18 Toucan (colour plate xviii)

C4-19 Dog

C4-20 Giraffe (colour plate xviii)

Coordinate Geometry

Coordinate geometry involves two lines drawn perpendicular to each other. The position of any point in the coordinate system can be given by a set of two numbers or letters that correspond to how the point is related to these two coordinate grid lines. In the popular children's game of Battleship, for example, a letter and number system are used to locate points on the grid (e.g., A4 means that a "torpedo" is being fired at the point positioned in row A, column 4). Note: Sometimes it is difficult for students to remember the difference between columns and rows, as defined in mathematics. One way to help them remember the difference between rows and columns is to think of the columns of a temple (like the Parthenon), which are always vertical. Hence, rows are horizontal, like rows of vegetables in a garden.

The commonly used Cartesian plane (named after French mathematician and philosopher René Descartes) divides the coordinate system into four quadrants using a vertical line and a horizontal line that intersect (i.e., the horizontal x-axis and the vertical y-axis). The four quadrants are numbered counter-clockwise beginning in the top-right quadrant. For example, the point (2, 3) means that the point corresponds to a distance of two units along the x-axis (or is situated two units away from the vertical y-axis) and corresponds to a distance of three units along the y-axis (or is situated three units away from the horizontal x-axis), which means the point is located in the first quadrant (I). Points in the Cartesian

C4-21 Battleship (Milton Bradley) game involves a letter/number grid pattern (colour plate xviii)

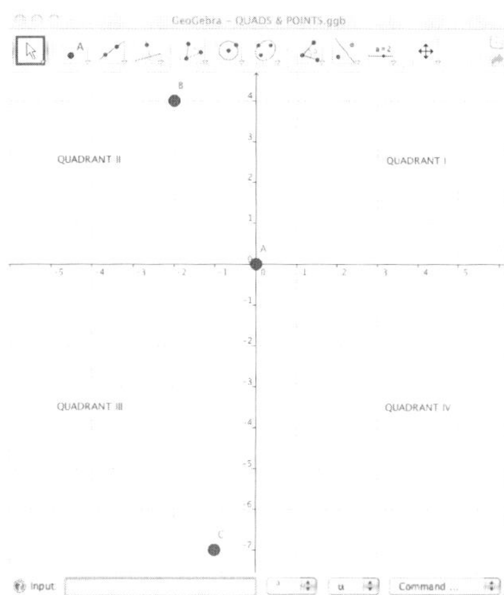

C4-22 Cartesian plane with three plotted points

plane system are always given with the horizontal coordinate first, followed by the vertical coordinate, so in the consistent form (x, y). The centre of the system is known as the "origin" and is located at (0, 0). Points located to the left of the origin have negative integer x-coordinates, and those below the origin have negative integer y-coordinates. So, the point (-2, 4) is located in the top-left quadrant (II), two units left of the origin and four units above the origin. The point (-1, -7) is located one unit left of the origin and seven units below the origin.

LEARNING EXPERIENCE
Animal Portraits Using a Coordinate System

▣ A coordinate system can be used to describe the face of an animal. Choose a full-frontal picture of an animal's face. Using a grid placed over the photograph (i.e., an acetate with a grid pattern photocopied onto it, or just regular graphing paper, perhaps taped to a window for added transparency), find the coordinates of the following points:

- Top of the head to the bottom of the chin
- Top of the head to the bridge of the nose
- Bridge of the nose to the bottom of the nose
- Bottom of the nose and chin
- Top and bottom of the mouth
- Corners of the mouth
- End points of the face across the mid-point of the eyes
- End points of the eyes (corners, top and bottom of widest part)
- Width of the bottom of the nose

Place these coordinates on a piece of graph paper. Using the coordinates as guides, draw the portrait. Then complete the portrait by adding detail, shading, hair, eyebrows, etc.

LEARNING EXPERIENCE
Mystery Animal Plot

- Using a Cartesian plane coordinate grid, draw or trace the contour of an animal that is positioned in such a way as to have at least some parts in each of the four quadrants. Starting at any point on the contour of the animal, have students establish points (dots) with (*x*, *y*) coordinates all along the contour. Have students then copy these coordinates in a list and exchange lists with fellow students, who must then try to re-establish and guess the animal by plotting and joining the Cartesian coordinates. In a group, discuss the following: Is there another or better way to find/plot coordinates? Where else might the origin be positioned on a grid system? Why do you think Descartes chose the middle of the grid for the origin in his coordinate system? Is it important to space the rows and columns evenly on the coordinate grid? Why? What are the benefits of using a single, international coordinate system, like the Cartesian plane system? What did you learn about art, drawing animal portraits, and mathematics in these activities?

LEARNING EXPERIENCE
Animals in Art and Design

- Pentagons and animals: Many animals have five appendages—four legs and a head. Like humans, they also have five senses and five openings in their head. Investigate the idea of pentagons (five-sided geometric shapes) and animals. Make a list of animals that have five appendages and/or paws with

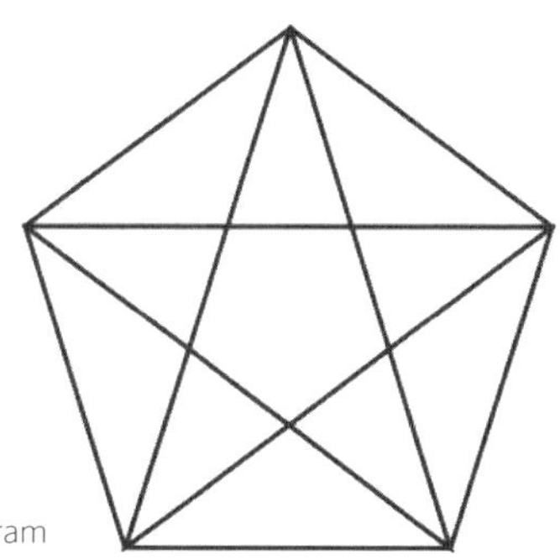

C4-23 Five-sided regular pentagon with inscribed pentagram

five toes. In mathematics, a regular pentagon denotes a five-sided shape with the special property that all five sides are equal in length. Have students find examples of regular pentagons in nature and in the designed environment.

- Plaster casts of footprints: Students can collect footprints of various animals, as well as their own foot- and handprints, by making plaster castings. Plaster castings from imagined animals can also be created in wet sand and plaster poured over to create the cast.
- Bilateral symmetry: Animal bodies, like human bodies, have bilateral symmetry. This is when the left and right half of the body are approximately the same, proportionately speaking. Collect pictures of frontal views of faces and bodies of animals. Cut the pictures in half vertically, and have students draw the mirror image of the remaining side.
- Animals in logos or motifs: Investigate animals and how they have been used as inspiration for elaboration and decoration. Students can study animals, their use in various cultures, and their history as symbols, including in motifs and logos. Students can create a personal or class logo, flag, or banner using coloured paper, paint, quilted or dyed cloth, etc.

Camouflage

Animals use several visual means as survival strategies: they attempt to hide or camouflage themselves in their environment, they impress other animals with their brilliant colours and shapes; and some animals and fish confuse their enemies, such as more dangerous animals or fish, by appearing to be what they are not.

Camouflage, or visual concealment, is a technique used by many living organisms for protection from predators. The idea of camouflage is also prevalent in human-designed objects such as vehicles (e.g., trucks, tanks, and airplanes) created and painted to blend into a particular background. When animals are visually camouflaged, their pattern scale and colour is similar to the surroundings.

LEARNING EXPERIENCE
Camouflage in Artwork

- Create an animal or other life form, real or imaginary, draw/construct the animal as a two-dimensional or three-dimensional form, and then conceal the creature in a replica or invented environment.
- Place bottles or other objects against various backgrounds and have students draw or paint the objects so that they "disappear." Similar exercises can be reversed with the new objective being to make the form stand out from its surroundings. These projects encourage students to observe various elements of art/design such as line, colour, texture, tone, and shape.

Sculpture and the Animal

The image of animals can be used in various sculptural projects using various media, including wire, papier mâché, clay, fabric and soft sculpture, and dioramas. Once students have observed and drawn an animal and become familiar with body proportion and movement, a wire sculpture of the animal could be the next project. Students will transfer what they know about movement and body proportion to produce a wire form. Consider the wire sculpture as a three-dimensional line drawing in space.

Line is an element of composition and any single line is, of course, related to many other lines when constructing a wire sculpture. There are numerous possible relationships. Repetition can give rhythm to a composition; variety in line length and thickness can also avoid monotony. Static or perpendicular lines relate to horizontal lines, which often bring stability to a composition. Dynamic lines, similar to the use of italics in writing, appear to be mobile, going in various directions.

LEARNING EXPERIENCE
Wire Sculpture Figure in Motion

- The materials needed are a variety of heavy, medium, and light wire. Supply several lengths for each student. Wires within the sculpture can be connected by twisting the wires around one another and/or by using a soldering gun or glue to make the connections more stable. The objective is for students to draw the figure in gesture using the wire. Sculptures can be painted or spray-painted outdoors. The wire sculpture can also become an armature for the application of papier mâché.

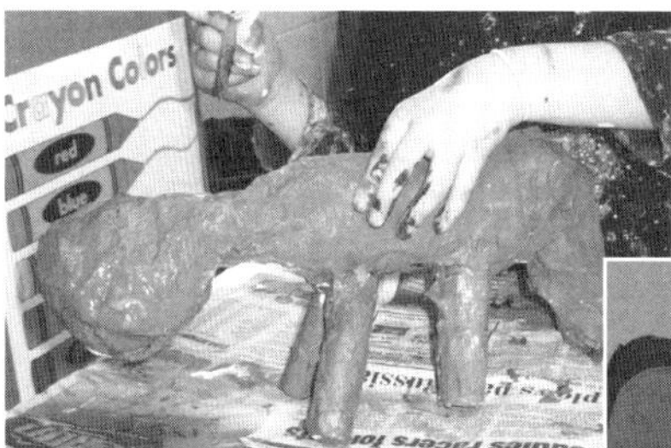

C4-24 Student creating an animal wire and papier mâché sculpture (colour plate xix)

C4-25 Animal wire and papier mâché sculptures (colour plate xix)

Tessellations

A tessellation is a repeated pattern using a geometric shape to completely cover a two-dimensional surface, without overlapping or leaving gaps between shapes (e.g., paving stones, wallpaper patterns, or ceiling tiles). One of the most prominent artists who extensively used tessellations in his work was M. C. Escher.

A "regular" tessellation means a tessellation that is made up of congruent, regular polygons. In other words, the shapes used in the tiling are "regular" (all sides and angles are equivalent) and "congruent" (all polygons are the same shape and size). Only certain polygon shapes work for regular tessellation patterning. To tessellate a plane, the interior angles of a regular polygon must divide evenly into 360°. Three shapes that are often used are squares, equilateral triangles, and hexagons. Semi-regular tessellations are formed by combining two or more different types of polygons. Tessellations can also be done on three-dimensional surfaces, such as the use of hexagons and pentagons to tessellate the surface of a soccer ball using the familiar black-and-white colour pattern.

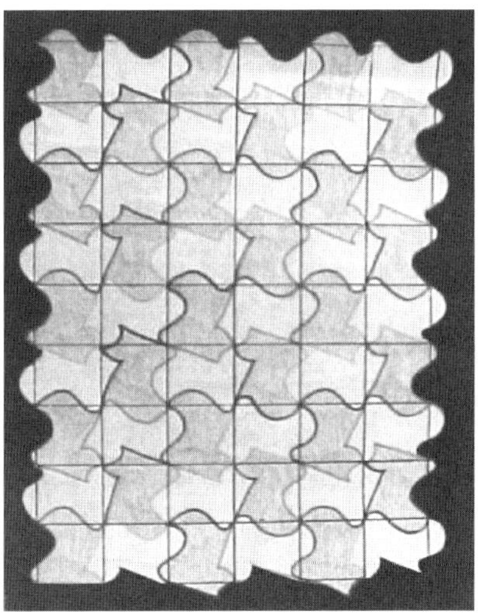

C4-26 Square-based tessellation pattern
(colour plate xix)

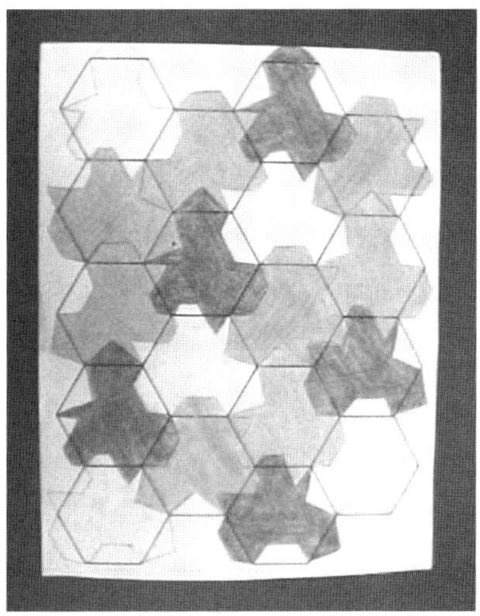

C4-27 Hexagon-based tessellation pattern
(colour plate xix)

LEARNING EXPERIENCE
Tessellation Investigation

◘ To tessellate a plane, the interior angle of a regular polygon must divide evenly into 360°. To determine mathematically whether various polygons may or may not be used to tile a two-dimensional surface, have students complete a blank version of this chart:

Regular Polygon	Number of Sides/ Angles	Measure of Interior Angles	Measure of Each Interior Angle	Will It Tessellate?
Trigon (triangle)	3	180°	60°	Yes
Tetragon (square)	4	360°	90°	Yes
Pentagon	5	540°	108°	No
Hexagon	6	720°	120°	Yes
Heptagon	7	900°	128.6°	No
Octagon	8	1080°	135°	No
Decagon	10	1440°	144°	No
Any "*n*-gon"	n	$(n - 2)\,180$	$\frac{(n - 2)\,180}{n}$	

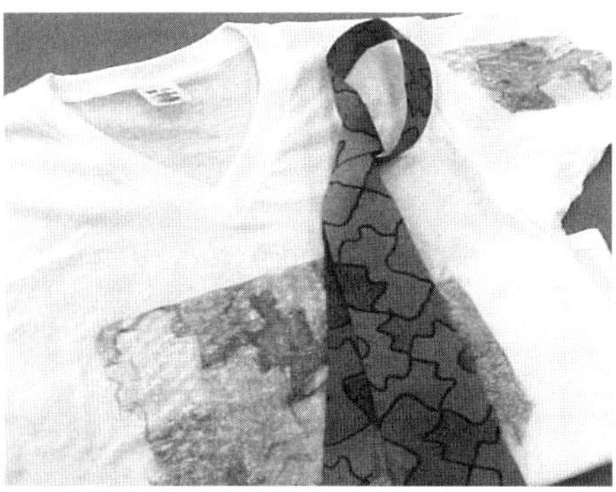
C4-28 Shirt and tie tessellations (colour plate xxi)

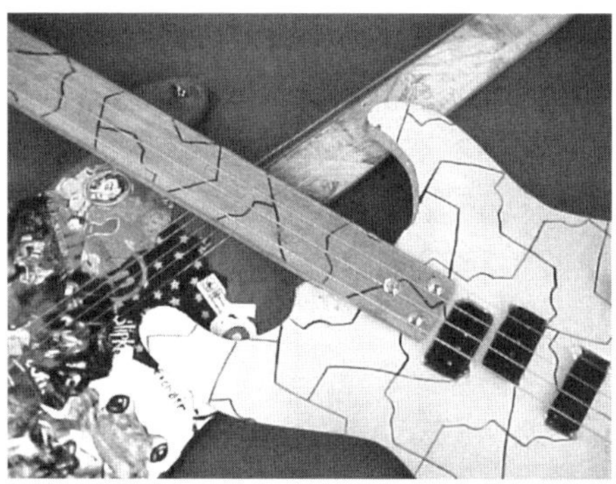
C4-29 Tessellated homemade guitars (colour plate xxi)

Conclusion: Only three regular polygons can tessellate a plane (three-, four-, and six-sided). Note that semi-regular tessellations can be formed by combining two or three polygons. For example, hexagons are used with pentagons on soccer balls.

LEARNING EXPERIENCE

Tessellations Involving Fauna Forms

C4-30 Creating a tessellation pattern (colour plate xx)

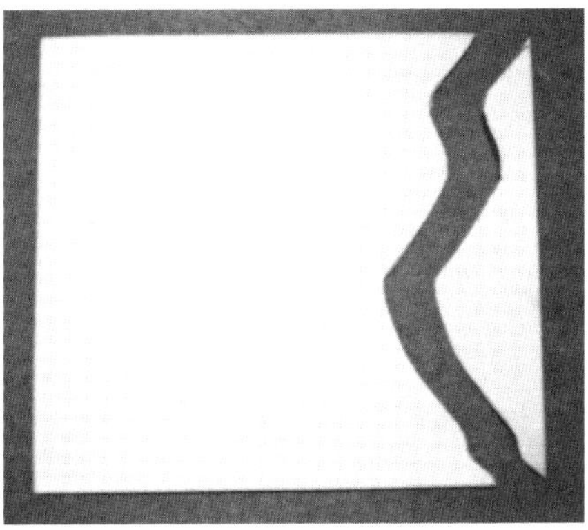

Step 1

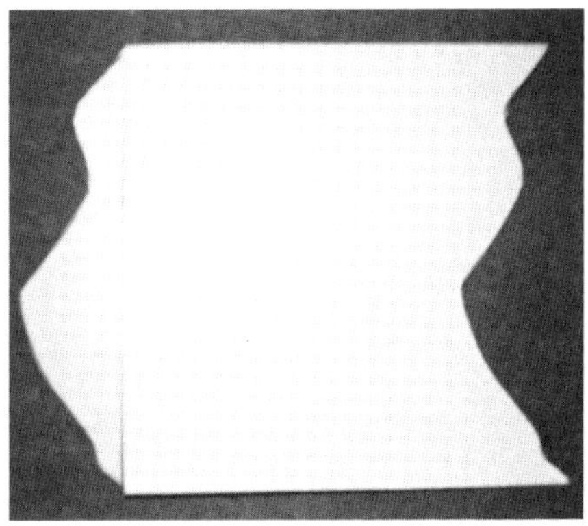

Step 2

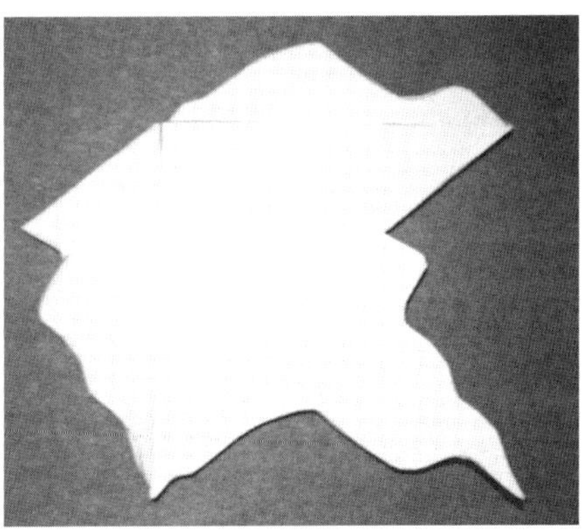

Step 3

Steps 4 & 5

1. Beginning with a square or a hexagon shape, students draw a creative line (i.e., this line can have straight, curved, wiggly parts but must start and end at the corners of the square template) from vertex to adjoining vertex on one edge of the shape. Students can create their lines for the template in such a way as to depict the head or body of a fish, bird, or land animal.
2. This line/shape is then cut out with scissors, slid across (i.e., translation) the shape template, and taped onto the opposite edge (see figure C4-30).
3. Steps 1 and 2 can be repeated twice for the square and three times for the hexagon, using different line/shape configurations for each of the opposite sides each time.
4. When the template is completed it can be traced in the middle of a sheet of blank paper, then slid in any of the four directions, and retraced again.
5. This tracing of the template is repeated in all four directions until the entire surface is tessellated (covered without overlap or gaps).

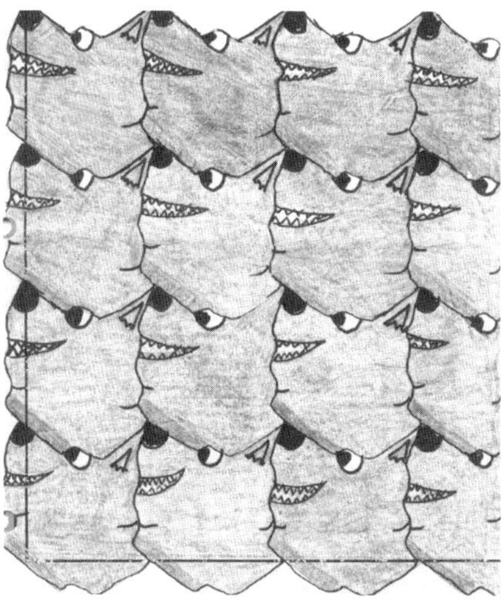

C4-31 Tessellated dog image (colour plate xxi)

C4-32 Aquatic life tessellation (colour plate xxi)

Refer to the works of Escher for ideas ranging from simple to complex. For example, during the template creation stage, instead of just sliding a drawn/cut shape across the template to the opposite edge and taping (translation), students can also rotate the shape and tape to an adjoining edge (rotation), or flip the shape, slide across and tape (reflection). These transformational moves (rotation, reflection) are then echoed in the tracing of the template: when sliding the template in any of the four directions to cover the surface, students will need to rotate or reflect the template to match the way in which is was constructed, so that all lines fit.

Aquatic Life

> *People have two legs, cats have four, insects have six and spiders have eight.... Starfish have five arms (or ten, eleven, even seventeen, depending on their species).* (Stewart, 1995, pp. 3–4)

It is estimated that oceans cover approximately seventy percent of the earth's total surface, which represents a remarkably vast area (361,132,000 km^2). Many cultures and civilizations, especially coastal groups, have been shaped by the seas or other large bodies of water.

Humans are often amazed by the sea, with its wide expanse of water, waves, and aquatic life forms, along with ever-changing and often highly dramatic weather conditions. Tsunamis (Japanese for "harbour wave"), or large tidal wave storms, can be severe and devastating, such as the one experienced in the Indian Ocean in 2004, which killed over 200,000 people in coastal villages and cities. The life of fish, birds, and people that live nearby or in the water provide endless sources of material for visual study. The following is a sketchbook or journal assignment based on sea life, but which can be adapted to almost any nature study.

LEARNING EXPERIENCE
Creating a Dossier of Aquatic Imagery, Text, and Video

- Compile a dossier of visual material about sea life (i.e., aquatic, marine life). Collect pictures or articles about sea life cut or photocopied from magazines. Collect shells, fish skeletons, lobster tails, and crab legs. Eat at a seafood restaurant and bring the remaining shells to class (after cleaning and boiling them). Make texture rubbings of these items and draw them using various media.
- Make an enlarged drawing of observed patterns on sea life or shells. A viewfinder is like a window or slide holder; it helps the artist focus on an area to study and draw. Make a viewfinder out of cardboard, hold it out and focus on a detail of a specific area.
- Find short stories or poems based on sea life or make up your own story or poem.
- Make a collage picture based on the theme of sea life (i.e., aquatic, marine life). Be especially concerned about the centres of interest, the overlapping of shapes, and the movement of line, colour, and shapes.
- Review newspapers to find line drawings that use various methods of suggesting form (e.g., dots, lines, cross-hatching).
- Search the Internet for video clips of marine life and/or sea-related human activities.

C4-33 Dolphin drawing

C4-34 Dolphin drawing using stippling technique

C4-35 Dolphin stencil design
(colour plate xxii)

C4-36 Fish potato relief print using sponge and other objects (colour plate xxii)

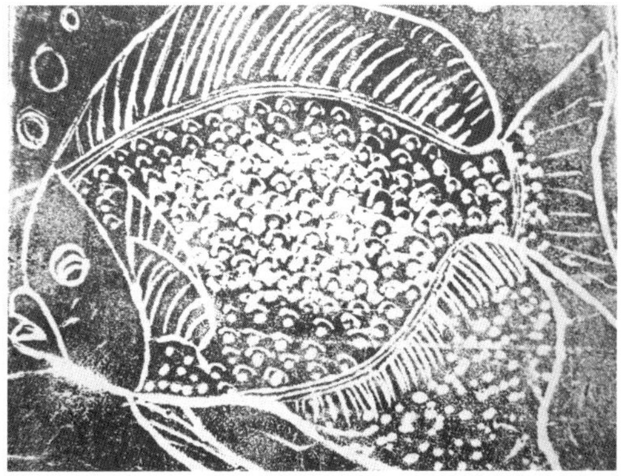

C4-37 Fish Styrofoam etching print

- Create a list of sea-life names or phrases that could be used as visual puns (e.g., sea-horse, hammer-head shark, star-fish, gold-eye, angel-fish, sun-fish, dog-fish, shark-fin soup, crab-legs, crab-net, es-car-go-t). Illustrate one of these creatively.

Amphibians and Reptiles

Amphibians include frogs, toads, and salamanders. Frogs begin life as aquatic tadpoles and undergo a radical transformation from birth to adulthood, allowing them to first breath through gills like fish, and then to later breath air with lungs as adult frogs. Whereas amphibians have smooth, moist, and sometimes rather sticky skin laden with mucous glands, reptiles have skin that is usually dry and scaly with small folds in the skin providing protection. Reptiles, including turtles, alligators, crocodiles, snakes, and lizards, have skin made up of scales that often features beautiful patterns of shapes and colours. Art, math, and science lessons could include the observation of pattern, colour, adaptation, camouflage, and the creature's environment.

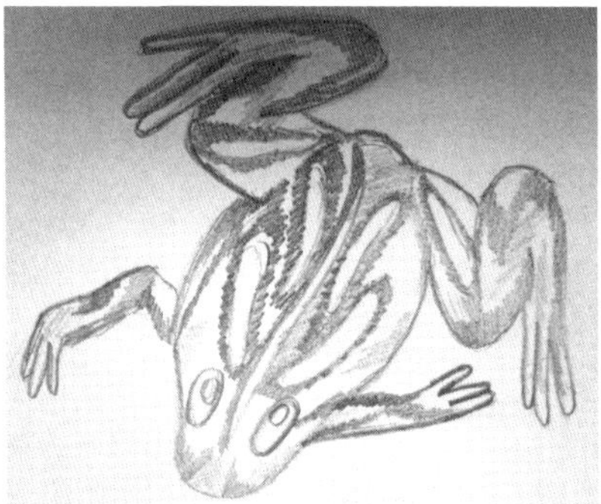

C4-38 Graphite frog drawing

C4-39 Frog colour drawing (colour plate xxii)

LEARNING EXPERIENCE
Drawing Dinosaurs and Sculpting Skeletons

Dinosaurs were very large reptiles that roamed the earth many years ago, but eventually became extinct. There are many areas where dinosaur skeletons have been found, excavated, and reassembled. These finds have given scientists and viewers some idea of their form and enormous size. Drumheller, Alberta, in Canada has a museum dedicated specifically to dinosaurs.

- Have students create detailed drawings of a favourite dinosaur from pictures. They can use the grid technique to increase the size of their drawing, or use a mathematical ratio (e.g., 1:10) to calculate the new lengths involved in the enlarged form.
- After drawing skeletons of dinosaurs or lizards, develop the skeletal structure of an animal you have invented. Is this animal a hybrid of two or more animals? Does it fly or swim, walk or crawl? Develop detailed features of this invented animal, such as body covering, feet, face, eyes, and ears. Describe its eating and behavioural patterns.

C4-40 Tyrannosaurus Rex dinosaur skull

C4-41 Dinosaur texture rubbings collage
(colour plate xxii)

Creatures of Myths and Legends

In Greek mythology, Icarus, ignoring the commands of his father, flew too close to the sun, and when the wax holding his wings melted, he plummeted to his death. Great eagles saved the protagonists from the point of certain doom in Tolkien's Hobbit tale. Fairies, unicorns, gryphons, phoenixes, and dragons—these and many other flying creatures have stirred the imagination of artists and writers for centuries. The study of cultural myths and legends can suggest an abundance of creatures of invention. This is a fertile and ideal area for curriculum connection, allowing teachers to incorporate social studies and language arts along with mathematics and the visual arts.

LEARNING EXPERIENCE
Creating an Original Myth/Legend Creature and Narrative

- Students can create their own sky creature (and/or land or sea creature, or combination thereof), along with a corresponding myth and legend. For some wonderful, imaginative examples of imaginary and mythological creatures, refer to the short television miniseries by Jim Henson (Muppets creator) entitled *The Storyteller*. These mythological creatures can be drawn, painted, sculpted, built, sewed, or formed using software. Myths/legends involving the new creature(s) can be written and/or acted out in class. Consider having students work in groups, combining their creatures in new stories.

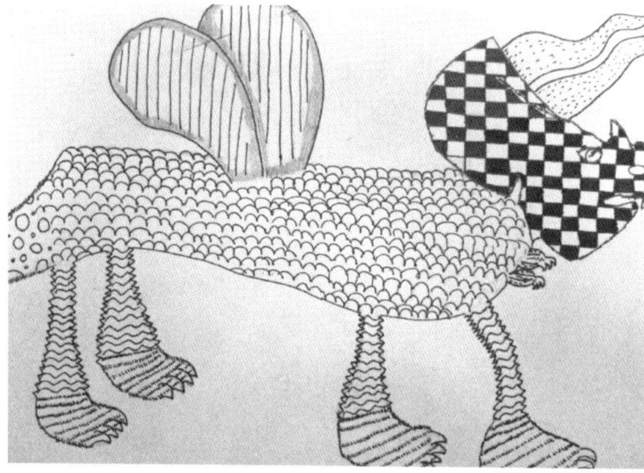

C4-42 Mythological creature design

Summary

In Chapter 4, we have seen how fauna of the land, sea, and skies have enriched the lives of humans since the beginning of recorded history. Whether they have focused on well-known, everyday creatures or on fantastical creations of the human imagination, students have been challenged to better understand our earth creatures and to create their own unique inventions. The power of modern hard- and software has allowed film directors to achieve incredible realism when it comes to rendering fauna in film. It is this combination of visual artistry and mathematical technology that allows Hollywood to capitalize on the incredible skills of artists and technicians to realistically render whatever can be dreamed up or imagined.

5 Mathematics and Art in the Human Figure

Protagoras (Greece, fifth century BC) proclaimed: "The measure of everything is human." There was an awareness of the importance of human measurement and a sense of harmony and symmetry. (Dilke, 1987, p. i)

Introduction

Throughout history, artists have measured people, objects, and distances found within their environment in order to understand visual depth, proportion, and distance for their art creations. Natural connections between art and mathematics appear to begin as people from early civilizations developed measurement systems based on the human body and body parts. For instance, they developed measurement systems to describe lengths and quantities by using body parts, such as fingers and hands, and they used body movements such as pacing to measure distances. The ten digits found on a pair of human hands played a significant role in the development of number systems. We count our ten fingers (and sometimes our toes). The shapes of human bodies and fingers also influenced the development of numbers and measurements. The Roman numeral five "V" was derived from the shape of the space between the thumb and the forefinger, and the ten symbol "X" is actually a combination of two V's (Schneider,

1994). The understanding of the human form and proportions greatly assisted the artist in working with naturally occurring and constructed objects.

Human Body and Measurement of Length

According to Johnson and Glenn (1961), "The breadth (width) of a finger was called a digit and the width of a human hand was called a hand or a palm. The distance from the end of the thumb to the end of the little finger when the hand was stretched out fully was called a span" (p. 7). The *cubit* was used by the ancient Babylonians, Sumerians, Hebrews, and Egyptians as the first recorded standard unit of measurement. Although the exact length differed somewhat among the above-mentioned cultures, the cubit was generally the length of a forearm from elbow to middle fingertip, or this same distance plus the width of the palm (Egyptian). To estimate distance, or the dimensions of larger spaces, pacing (regular foot strides) was often used.

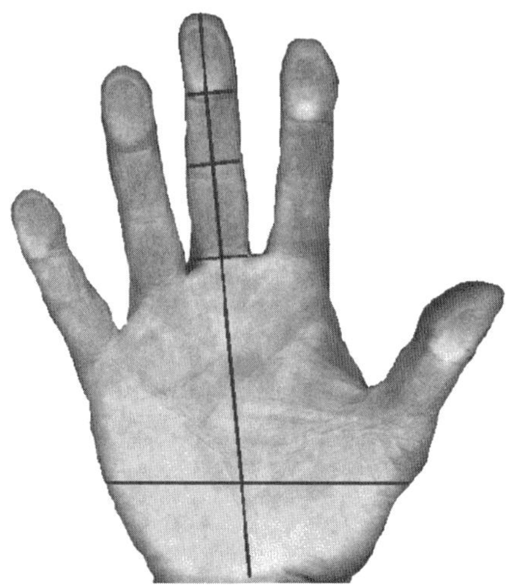

C5-1 Human hand measurement

The *Olympic cubit* was used by the early Greeks and Romans and two-thirds of this measurement became what western cultures call a *foot*, or 12 inches. The *yard* was devised by cloth merchants as the distance from the middle of the chest to the fingertips of an outstretched arm. King Henry I of England (1068–1135) introduced and enforced the *standard yard* (3 feet) for measurement, which was the distance from the point of his nose to the end of his thumb.

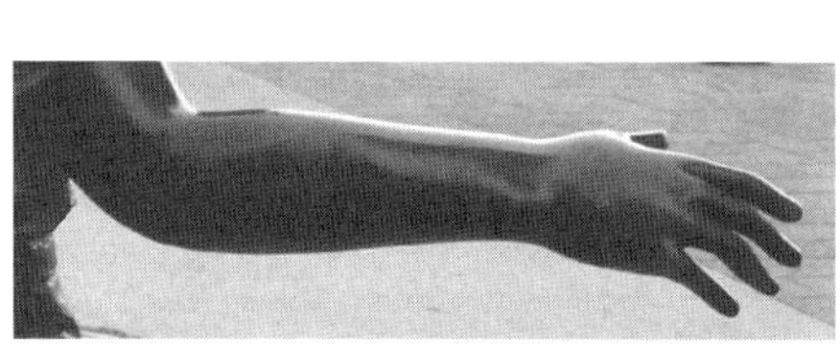

C5-2 The cubit was measured by the length of a man's forearm (18 inches/46 cm)

As societies developed, an increase in trade, building, and other construction activities precipitated the need for standards of measurement. Metric measurements are now used in Canada, England, and many other countries in an effort to standardize the trading economy and make measurement and quantities more consistent.

LEARNING EXPERIENCE
Human Body Measurement

- Create a table to record measurements for four or five people. Some measurements that can be recorded are the digit (finger), palm (hand), arm span, foot length, and pacing distance. As a class, collect results and compare and graph them.

	Digit	Palm	Arm Span	Foot	Pacing Distance
Person 1					
Person 2					
Person 3					
Person 4					
Person 5					

LEARNING EXPERIENCE
Measuring and Graphing Human Body Parts

- Measuring body parts with string or yarn increases students' awareness of size and proportion. Have pairs of students take a piece of string and measure their various body parts, such as feet, head, hands, forearms and legs. Also have them measure parts of the head such as the nose, lips, and ears. Students can lay out their strands of string and look for comparisons between their own string lengths and those of their classmates. "Are there patterns that are readily evident in [string] lengths and parts of the body?" (Kohl & Gainer, 1996, p. 226). Students can measure these patterns and create a graph or scatter plot from the data.

- For an art-making experience, students can re-create their own body using white glue and string lengths for the various parts, gluing the strings onto sheets torn from large rolls of paper. Students can add details to their figure, such as fingers, facial features, clothes, or shoes. As a variation to this exercise, have students trace each other on a large piece of paper, measuring and labelling different parts of their bodies (Ritter, 1995).

Proportion and the Human Figure

Vitruvius (Marcus Vitruvius Pollio, 31 BC–AD 14), a Roman architect and engineer, wrote that a well-built man is symmetrical, with parts that are in proportion. He studied the proportions of the human body and discovered that the face, measured from the chin to the hairline, was one-tenth the length of the body. At the approximate centre of the body was the navel (Verstockt, 1987). He wrote, "Nature has thus arranged the measurement of man: four fingers make one palm; four palms make one foot; six palms make one cubit; four cubits make once a man's height."

Michelangelo (Michelangiolo Buonarroti, 1475–1564) made countless drawings and measurements of the human body (both living and dead subjects) as well as of existing statues, and he found that, although the measurements varied, the classical "ideal" proportion of the human body was eight "heads" tall. However, he created his famous statue *David* seven heads tall. This may have been to help give the effect that David was still a youth.

Leonardo da Vinci (1452–1519) was also very interested in the proportions of the human body, and his contemporary, Albrecht Dürer (1471–1528), wrote specifically on the topic of proportion in his work *Vier Bücher von Menschlicher Proportion* (1528). Dürer favoured seven-and-a-half head lengths for the body proportion. The Dutch painter Rembrandt Harmenszoon van Rijn (1606–1669) was reported to have also used seven. Alberto Giacometti (1901–1966), an artist from Switzerland, at times used 12 head lengths. According to Edmund Burke (1909), "Some hold it to be seven heads: some make it eight; whilst others extend it even to ten; a vast difference in such a small number of divisions" (p. 2).

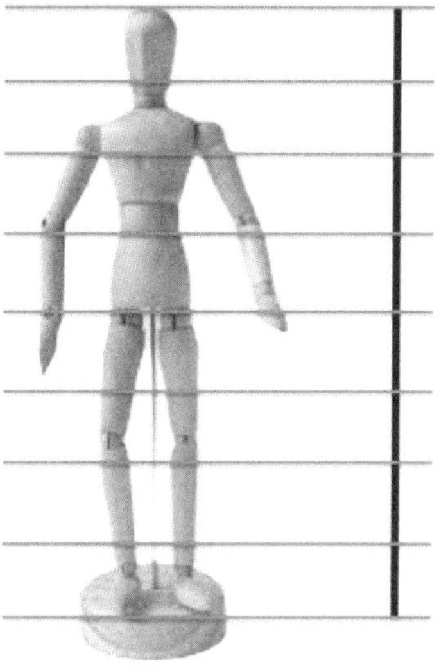

C5-3 Human proportion (eight heads tall)

LEARNING EXPERIENCE
Human Body Ratios

1. Create a chart with at least five rows and eight columns.
2. Have students measure their head (from top to chin), hands, and feet, as well as their overall height. Students could also make a photocopy of a Vitruvius sculpture (or another artist's male or female sculpture) in a standing pose, make similar measurements, and include these ratios on the chart.
3. Calculate the head-to-body ratio.
4. Calculate the foot-to-body ratio to determine how many "feet" tall the person is.
5. Calculate the hand-to-body height.
6. Once students gain an understanding of proportion, set up a mannequin, doll, skeleton, or live model for students to draw in gesture, mass, or contour.

	Height (h)	Head	Hand	Foot	Head/h	Hand/h	Foot/h
Person 1							
Person 2							
Person 3							
Person 4							

LEARNING EXPERIENCE
Who Fits in a Square?

Leonardo da Vinci developed images showing the bilateral arrangement of the body in a circle and in a square. He established the geometric relationship between the square and the circle, and the geometrical proportions of the human body (Verstockt, 1987, p. 114). One famous image, known as Vitruvian Man, has been used by organizations in many advertising and marketing campaigns for various products and events.

- Have students measure their height and arm span (fingertip to fingertip). Are these measurements equal? Use these measurements to create four points on the floor or wall. Do these four points indicate the four points of a square, circle, or rectangle? Are the arm span lengths equal to the height? Gather measurements (length of arm span and height) from other students and create a chart with the class data. Do most people have a longer arm span than height, or are these measurements about equal?
- This experience can help students gain a better understanding of human proportion and can be followed with a drawing or sculpture assignment. Students can increase the reliability of their data by expanding the number of people measured, including friends and family members outside of the school context, and reporting back results to the class for group analysis.

C1-1 Iris images repeated in various media

C1-2 Flowers with five petals

C1-5 West Coast totem sculpture of human figure

C1-4 First Nations teepee

C1-6 Greek Parthenon

C1-10B Painting the train

C1-9 The joy of making lines and images as a child

C1-7 Raphael's *The School of Athens*

C2-1 Manipulatives and problem-based learning in mathematics education

C2-2 Fourfold object design

C2-3 Line

C2-4 Shape

C2-5 Texture

C2-6 Space (Form)

C2-9 Balance

C2-7 Value (Tone)

C2-8 Colour

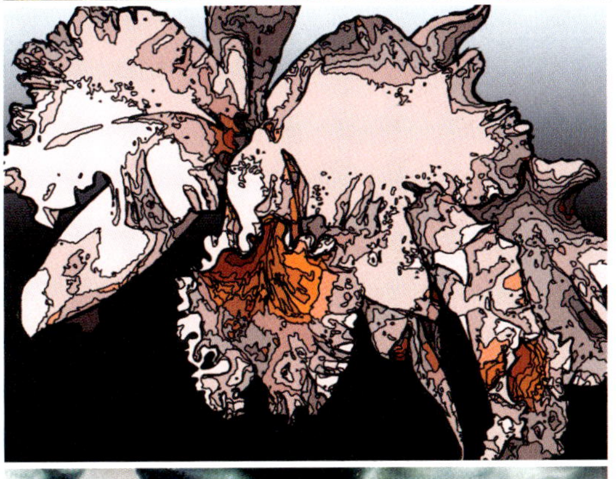

C2-10 Contrast (Variety)

C2-11 Emphasis (Dominance)

C2-12 Pattern (Repetition)

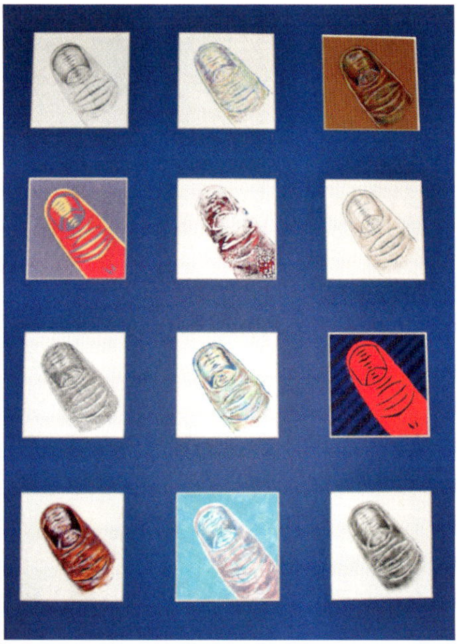

C2-13 Rhythm (Movement)

C2-14 Proportion

C2-15 Unity (Harmony)

C2-16 Printmaking with stamp and colour ink

C2-17 Watercolour painting

C3-1 Exploring the natural world

C3-3 Examining Flowers

C2-18B Kiln oven for clay firing

C3-2 Corn cob patterns

C3-4 Observing/ recording tomato plant growth

C3-5 Relief stamp print of tree

C3-6: Apple with star pattern and golden ratio

C3-7 Daisy

C3-8 Romanesco with radial symmetry pattern

C3-9 Fiddleheads with spiral growth pattern

C3-10 Shell with spiral pattern

C3-11 Sunflower seed patterns

C3-12 Spiral plant pattern

C3-14 Fern fractal patterns

C3-18 Beebalm flower detail for macro drawing

C3-16 Shape It, a game of visual perception

C3-19 Using a viewfinder made of two L-shaped mat board pieces

C3-20 Using your hands as a portable viewfinder

C3-21 Community quilt with blocks depicting local cultural scenes and symbols

C3-27 Naturally occurring geometric tiling pattern in honeycomb

C3-28 Leaf tessellation pattern

C3-29 Kaleidoscopic flower designs

C3-30 Fall tree colours in North Bay, Ontario, Canada

C3-32 Tree painting

C3-33 Tree leaves watercolour painting

C3-34 Dendrochronology

C3-38 Maple leaf vein patterns

C3-40 Pine cone scale patterning

C3-41 Landscape painting with foreground, middle ground, and background

C3-43 The Golden Rectangle featuring "whirling" square pattern

C3-42 Landscape photograph of trees and lake

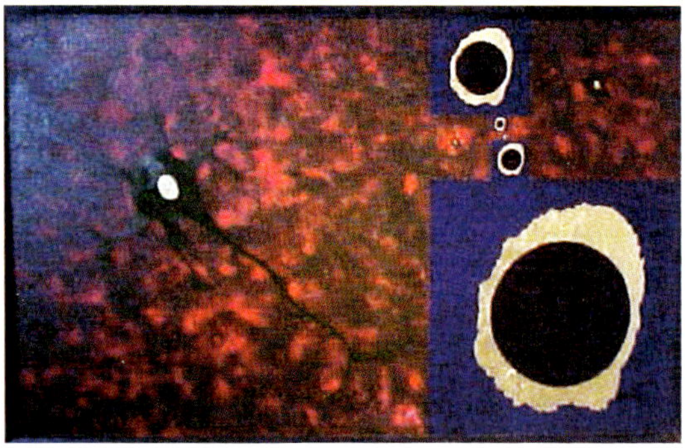

C3-44 Golden rectangle painting with iterating neural dendrites (rise) and solar eclipse (fall) images

C3-45 Rainbow colours

C3-46 Clouds and Great Lakes vista on Lake Huron near Goderich, Ontario, Canada

C4-1 Tiger head symmetry, positive and negative balance

C4-2 Butterflies

C4-4 Large snapping turtle

C4-9 Elephant riding in Chiang Mai, Thailand

C4-13
Penguin
colour
drawing

C4-18 Toucan

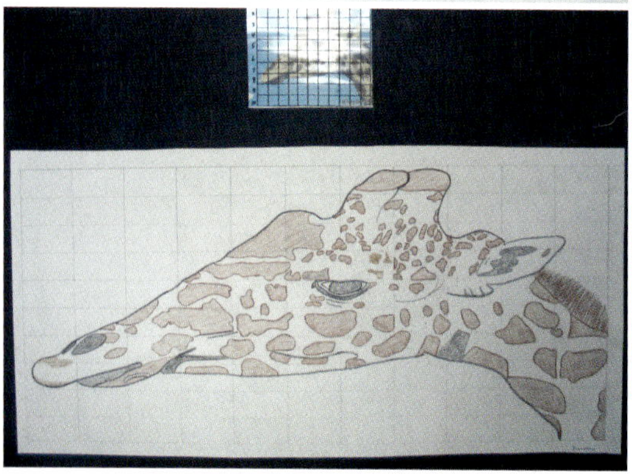

C4-20 Giraffe

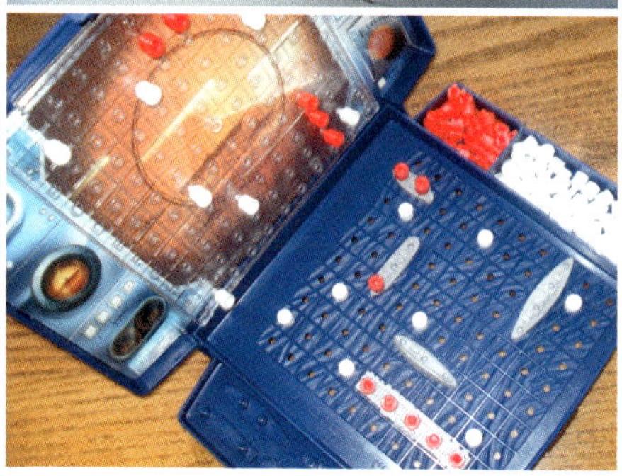

C4-21 Battleship (Milton Bradley) game
involves a letter/number grid pattern

C4-24 Student creating an animal wire and papier mâché sculpture

C4-25 Animal wire and papier mâché sculptures

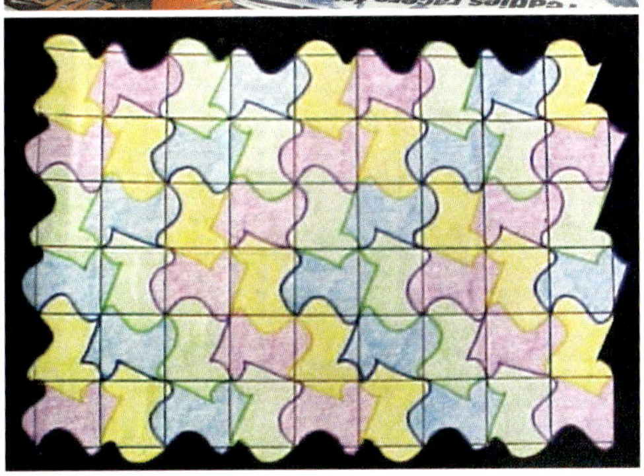

C4-26 Square-based tessellation pattern

C4-27 Hexagon-based tessellation pattern

C4-30 Creating a Tessellation Pattern

Step 1

Step 2

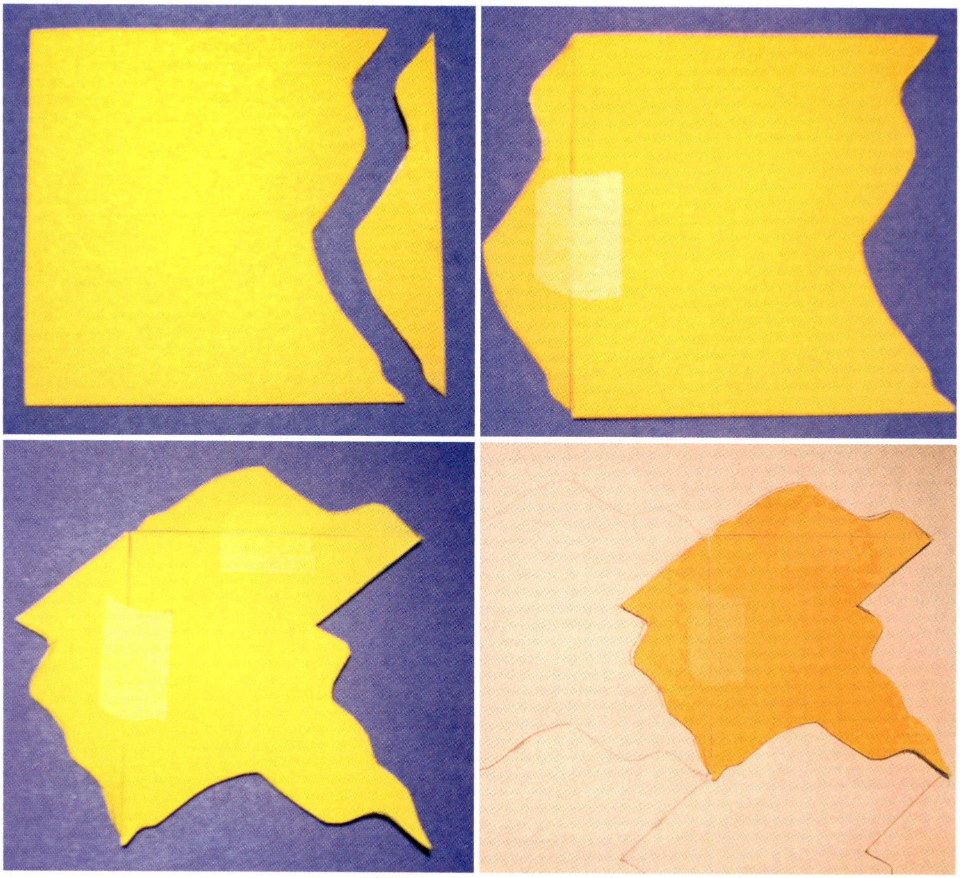

Step 3

Step 4

C4-28 Shirt and tie tessellations

C4-29 Tessellated homemade guitars

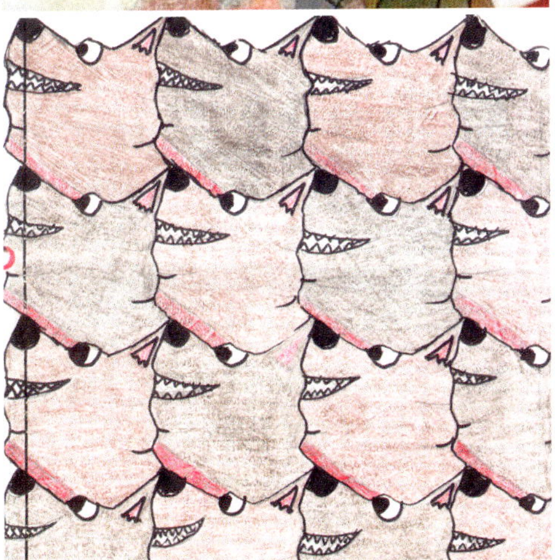

C4-31 Tessellated dog image

C4-32 Aquatic life tessellation

C4-35 Dolphin colour stencil design

C4-36 Fish potato relief print using sponge and other objects

C4-41 Dinosaur texture rubbings collage

C4-39 Frog colour drawing

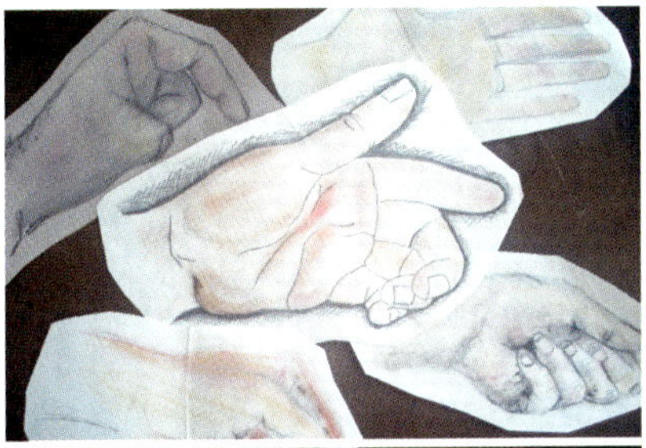

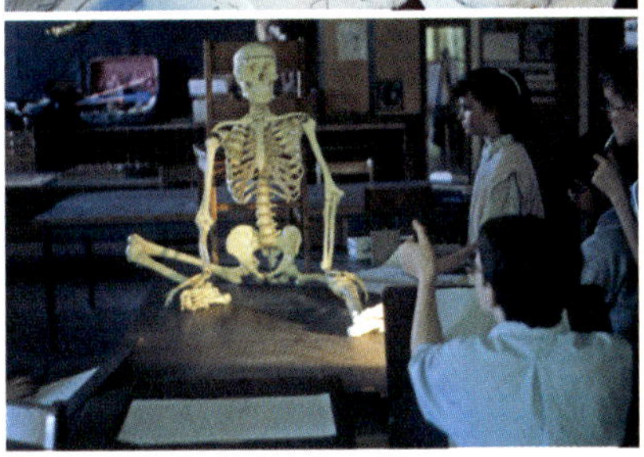

C5-21 Observing the human skeleton in drawing class

Top of page: C5-7 Fingers and spaces: student hand drawings with graphite and chalk pastels. Middle: C5-17 Instructor samples of direct measurement for human portraits

C5-23 Wire action figure sculptures

C5-25 People box sculptures with found objects

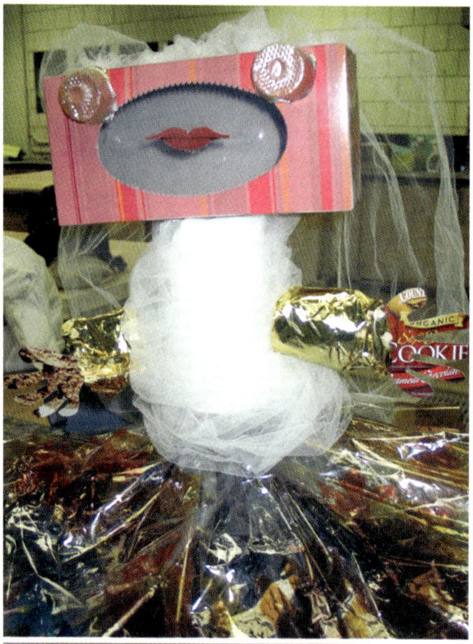

C5-24 Papier mâché life-size sculpture of the victim in Goya's *The Third of May* (1808)

C6-4 Circular design of the London Eye Ferris wheel in London, UK

C6-1 Westminster Abbey façade in London, UK

C6-5 Triangle design in roof trusses provides structural support

C6-6 Triangle design in wooden supports for Newton's mathematical bridge in Cambridge, UK

C6-7 Rectangular designs in the complex façade of Notre Dame Cathedral in Paris, France

C6-8 Pentagon design of the Pentagon in Washington, DC

C6-9 Hexagon design in table umbrella

C6-10 Octagon design in picnic table

C6-11 Hexadecagon design in outdoor pool

C6-12 Dome design of Inuit igloo constructed with snow and ice

C6-13 Dome design of the Dome of the Rock Islamic Mosque on Temple Mount in Jerusalem

C6-14 Egyptian pyramids

C6-15 Glass pyramid of the Louvre Museum in Paris, France

C6-17 Intricate stone tracery designs, Barcelona, Spain

C6-16 Greek column styles: Doric, Ionic, and Corinthian

C6-18 Living wall at the University of Waterloo Summit Centre for the Environment, Huntsville, Ontario, Canada

C6-19 Jewish Museum by Libeskind in Berlin, Germany

C6-20 Parthenon three-dimensional model structure

C6-26 Streets and city planning

C6-28 Centre Street Bridge over the Bow River in Calgary, Alberta, Canada

C6-29 Lion's Gate on Chain Bridge over the Danube River in Budapest, Hungary

C6-30 Tower Bridge over Thames River in London, UK

C7-1 Colour façade of the W. C. Cameron building in Victoria, British Columbia, Canada

C7-2 Christ Church Cathedral interior space in Fredericton, New Brunswick, Canada

C7-19 Student mandala design

C7-21 Intricate egg painting designs using wax resist and dyes

C7-22 Big Ben Clock Tower of British Parliament in London, UK

C8-1A Golden Rice Krispies square tower

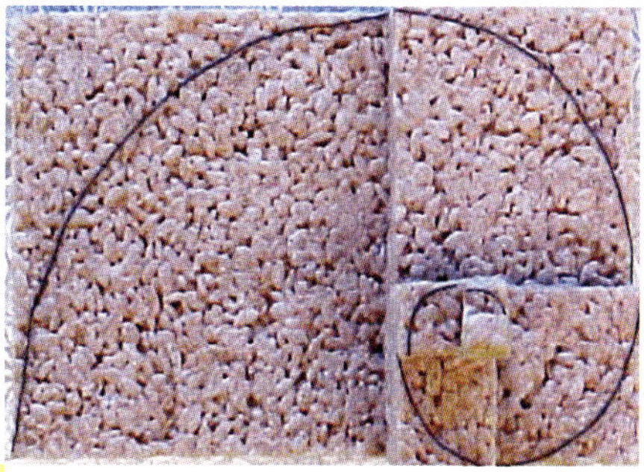

C8-1C Planets arranged by descending sizes of circumference

C8-1D Golden spaghetti noodles, meatballs, and fork tines

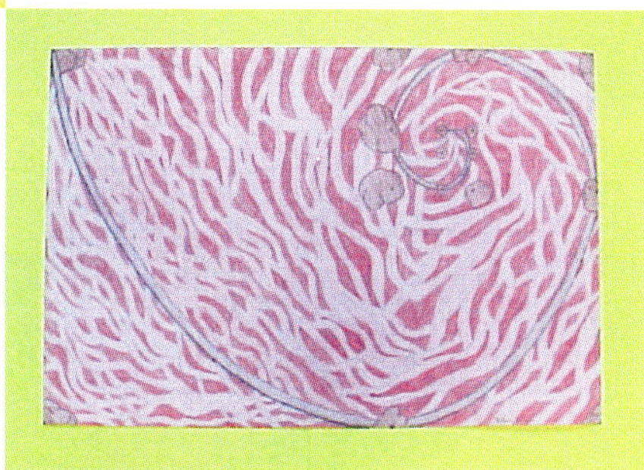

C8-1F Golden shapes and vocabulary

C8-2A Calculus project: Alison's "Adventure through the Calculus" illustrated storybook

C8-2B Calculus project: Mobile featuring two creators of the Calculus (Newton & Leibniz)

C8-2C Calculus project: Newton poster illustrating his quotation about finding pebbles on the seashore

C8-2D Calculus project: Monopoly—The Calculus edition

	Height (h)	Arm Span (a)	Difference (h - a)
Person 1			
Person 2			
Person 3			
Person 4			

LEARNING EXPERIENCE
Exploring Scatter Plots, Linear Regression, and Lines of Best Fit

A scatter plot can also be created with the same data set, plotting the heights of students on the horizontal axis against the corresponding arm spans (i.e., distance between fingertips when arms are outstretched) of students on the vertical axis. Linear regression is a measurement in mathematics of how well two sets of data relate to one another. It can be explored visually with the scatter plot, or algebraically with linear regression software or hand-held graphing calculators. In looking at the class data set that has been carefully recorded on the scatter plot, imagine a straight line that, when placed on the graph, would best separate the data points into two equal halves or sets of points. This is an approximation of the "line of best fit." The more the data points appear to cluster around a straight line, the better the "fit" or relationship between the two sets of numbers. So, in this case, if the data points all tend to cluster around a line, then we can assume that height is related to arm span. If the data points are spread out randomly around the graph, then there is unlikely a strong relationship between the two sets of values. Using graphing calculators or computer software (e.g., Texas Instruments TI-83 calculators, GeoGebra), students can find the actual line of best fit and the linear regression equation for their data set. This equation will feature something called a correlation co-efficient denoted by the letter *r*. The closer the coefficient

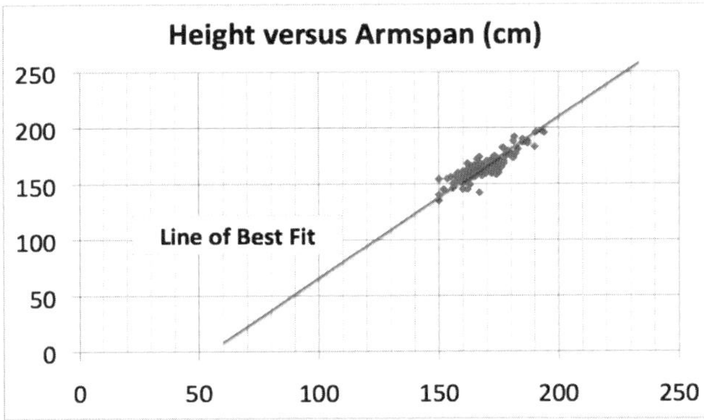

C5-4 Height versus Arm Span (cm) Showing Line of Best Fit

r is to the number 1, the closer the data sets are to having a perfectly linear relationship, i.e., foot length is directly/exactly proportional to human height. The closer to a zero value for *r*, the weaker the relationship.

The Golden Ratio in the Human Body

In mathematics, a ratio is defined as a comparison of two numbers. A proportion is an equality between two ratios (for example, a/b = c/d). Perhaps the most well-known ratio in human history is that known as the golden mean, or golden ratio (Livio, 2002; Runion, 1990). In simplest terms, the golden ratio can be defined as a situation in which the shorter section of a divided line segment compared to the longer section is in the same proportion as the longer section compared to the entire line segment. In other words, "short is to long as long is to whole." On any given line segment, this proportion exists only at one particular spot along the line. Mathematically, this special ratio can be easily derived using the "mean and extreme ratio" definition in mathematics (i.e., "cross-multiplying" to solve for the unknown variable, which in this case is an irrational, non-repeating decimal of approximately 1.61803) and the quadratic function (Jarvis, 2002; 2007).

The golden ratio is also related to the Fibonacci number sequence of 1, 1, 2, 3, 5, 8, etc., which is produced by adding two consecutive numbers, beginning with 1 and 1, to arrive at the next number in the sequence (so, 1+1=2, then 1+2=3, then 2+3=5, and so on). By dividing successive numbers in the sequence (e.g., 1/1=1, 2/1=2, 3/2=1.5, 5/3=1.66666, 8/5=1.6, etc.), the quotients become closer and closer (i.e., from above and below) to a special value, or limit, which has been designated by mathematicians with the Greek letter phi or 1.61803. Even more amazing than this fact, *any two* intergers (even negative integers) can be used to begin a Fibonacci-type sequence (e.g., 2, 7, 9, 16, 25, etc., or -7, 2, -5, -3, -8, etc.) that will approach phi (1.61803) when successive pairs of numbers are divided and the quotients plotted on a grid (see C5-6).

This golden ratio can be found throughout nature in plants and shells (Newman & Boles, 1992), and many have argued that this ratio even occurs within the human body, arm, and hand. For example, the Greeks would measure from the tip of the head to the floor and then divide this by the distance measured from the navel to the floor to determine the degree to which one's proportions were "ideal" or close to the golden ratio (approximately 1.61803).

1	1
1	2
2	1.5
3	1.666666667
5	1.6
8	1.625
13	1.615384615
21	1.619047619
34	1.617647059
55	1.618181818

2	3.5
7	1.285714286
9	1.777777778
16	1.5625
25	1.64
41	1.609756098
66	1.621212121

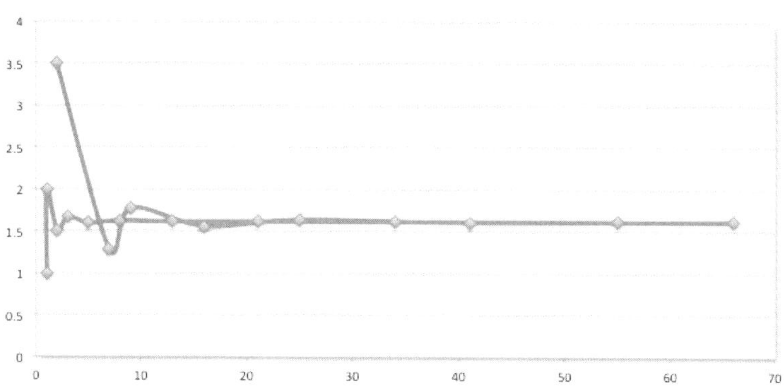

C5-5 Line Graph Showing Calculated Quotients of Two Sample Sequences Converging at Phi (1.61803)

LEARNING EXPERIENCE

The Golden Ratio in the Human Body

1. Measure one's full height from floor to top of head (*a*), then measure from the floor to one's navel (*b*). Divide *a*/*b* to compare with the golden ratio 1.61803.
2. Measure the arm from the shoulder to the tip of the hand's middle finger (*a*), then measure from the shoulder to the elbow (*b*). Divide *a*/*b* to compare with the golden ratio 1.61803.
3. Measure the middle finger from knuckle to the tip (*a*), then measure from the knuckle to the first joint (*b*). Divide *a*/*b* to compare with the golden ratio 1.61803.
4. Have students work in pairs and collect this data for the entire class.
5. With grid paper or a simple spreadsheet application, have students create graphs of this data for comparison purposes. Do some measurements come close to the golden ratio of 1.61803? How does this understanding of the golden ratio assist the artist in creating two- and three-dimensional artworks based on the human figure?

C5-6 The violin contains the golden proportion in terms of the ratio of its main parts.

Learning to See and to Draw

The brain is divided into two main hemispheres that are connected in the centre by the corpus callosum. The exact functions of these hemispheres in daily thinking and activity has been the topic of much debate for several decades, particularly in light of recent advancements in brain research and related technologies. In 1981, psychobiologist Roger Sperry won the Nobel prize in Physiology/Medicine for his work on functional specialization of cerebral hemispheres in the brain, and many other neuroscientists have continued to research in this area over the past three decades.

In her popular art education text, *The New Drawing on the Right Side of the Brain*, Edwards (1999) based her teaching and learning strategies for realism drawing on this distinction between the right and left hemisphere functions of the human brain. She suggested that beginning artists gain an insight into their drawing ability by forcing the right brain, the assumed region of artistic endeavour and creative thought, to become active in a series of carefully-designed drawing exercises. For example, Edwards has her students draw a reflected profile of a human face, wherein the well-recognized parts of the face become abstract shapes in reflection, forcing the right brain to be involved (pp. 46–50). When the tops and bottoms of the profile and reflection are joined by straight lines, the resulting drawn object resembles a flower vase. According to Edwards, right-brain thinkers have a greater sense of spatial relations. Edwards contends that all students of art can train their brain to be more spatially aware of line and shape by simply practising these types of shape copying exercises.

LEARNING EXPERIENCE
Profile Shadow and Reflection

- Another alternative to Edwards' face-vase drawing exercise is to trace a human face by using a strong light source or overhead projector in order to create a shadow of the profile. Have students trace this facial profile on paper, stopping at a line that is drawn vertically from the top of the head down through the ears to the neckline. Then have students reflect the traced line across the vertical/central line of symmetry and carefully draw the perceived reflection of the profile. Students can then cut out this profile and trace the reverse or mirror image to see how accurately they have rendered the reflection. Through repetition of this activity using different profiles or other traced shapes, students become increasingly adept at seeing spaces and relations through close observation instead of drawing from memory.

- To further extend and enrich this activity, students can manipulate the images to play with the positive shapes and negative spaces, cutting and reshaping the profile to create patterns and using various colour schemes to create images that emphasize the principles of composition, such as balance, unity, emphasis, or pattern.

The Fingers and the Spaces

The following learning experience is designed to overcome poor self-confidence in one's ability to draw (an "art phobia" situation perhaps, much akin to math phobia or anxiety). But first, a significant word of explanation. We do not, as co-authors, mean to convey a sense that realism drawing is the climax of artistic endeavour, or that it is to be valued above any other area of the visual art curriculum. With the many and varied branches of contemporary visual art—itself positioned within an even greater variety of expressional forms to be found within the fine arts (i.e., music, dance, drama, visual art)—realism drawing is admittedly but one part of a much greater whole. That being said, we contend that it is an important part, and sadly an often-ignored part in today's art curriculum. If, as Edwards contends, young teenagers commonly reach a point in their artistic development where they desperately wish to be able to draw something accurately (even within an age of digital photorealism), and if this desire is road-blocked by lack of instruction and practice at that developmental junction, then it is little wonder why children become frustrated with visual art and abandon their "stickmen" drawings (and perhaps artistic/creative pursuits in general) for the rest of their lives.

If adolescents were never taught the basic four operations in mathematics (add/subtract/multiply/divide), they would find it very hard indeed to (a) appreciate the historical significance of these concepts/skills in the course of human history and achievement (mathematical/scientific/technological endeavours), and (b) pursue any further studies in fractions, algebra, calculus, etc. without this knowledge.

This learning experience proves to students (young or adult—any beginning art student) that they do actually have the ability to draw when their focus is intentionally concentrated on smaller portions or parts of the whole image, rather than trying to focus on the entire image all at once. Many children and adults have developed a complete lack of confidence in their ability to draw realistically, especially when dealing with difficult subject matter, such as portraits or the human figure (i.e., as opposed to trees or animals, which are arguably less demanding in terms of detailed drawing). Edwards (1999, pp. 50–55) developed a technique that helped her art students to overcome this lack of confidence by having them reproduce a line drawing that is given to them upside down. Once the drawing is completed, the image and the drawing are then turned 180° for students to see what they are drawing. Edwards maintains that, almost without exception, her students are consistently amazed at their own drawing ability as evidenced through this exercise.

LEARNING EXPERIENCE
The Fingers and the Spaces

- Have students hold out their non-writing hand and repeat the following phrase: *"The fingers and the spaces are both important places."* As silly or simple as this may sound, within this phrase is the real key to high-realism drawing. Which spaces does the brain treat with more value or importance—the fingers themselves or the negative spaces in between the fingers? In drawing, these spaces are of equal value. However, because of the way the mind works, it wants to value the fingers/foreground objects more. Being aware of negative spaces is critical for realism drawing. You may note that when asked to draw their opposite hand, students will often fixate on the fingers, erasing contour and inner detail lines over and over again, not able to make it "look right." In this situation, tell students to force their eye to move one space over and draw the negative shape between two adjacent fingers. Then they can go back again to work on the fingers, then move over to the next adjacent space, and so on until the entire hand is rendered. If students draw accurate negative spaces, they will end up with perfect (or very well-rendered) fingers.

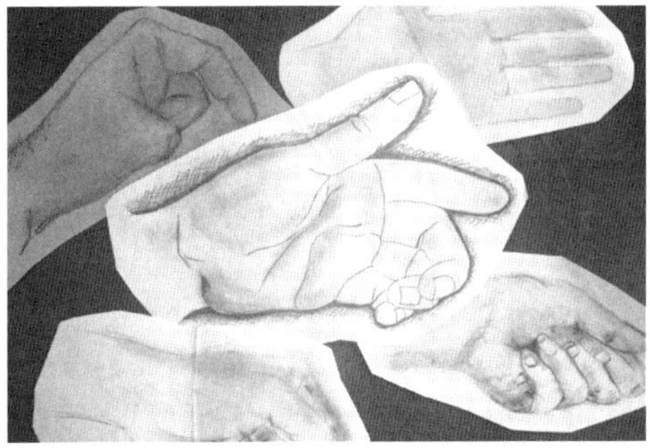

C5-7 Fingers and spaces: student hand drawings with graphite and chalk pastels (colour plate xxiii)

When we later arrive at human portraiture, perhaps the most difficult of all realism drawing exercises (a difficulty likely due to the incredible, inherent ability of humans to recognize familiar faces), it is this same, honed ability to "see" and to draw the whites of the eyes (i.e., the negative space next to the colourful iris) that allows the artist to accurately capture the facial features of the sitter/photo. Practice makes . . . better and more confident art students.

Drawing Body Parts

The Hand

The hand, with its four fingers and opposable thumb, helps humans grasp and hold objects. As mentioned in the chapter introduction, the hand is also an important object in the development of numbers and measurement. The skeletal structure of the hand is composed of bones arranged into two groups: the bones of the palm (metacarpals), and the bones of the fingers (phalanges). Each finger is attached to a metacarpal and is made up of three small bones: the proximal (closest to the palm), the middle, and the distal (at the end of the finger) (Marieb, 1987).

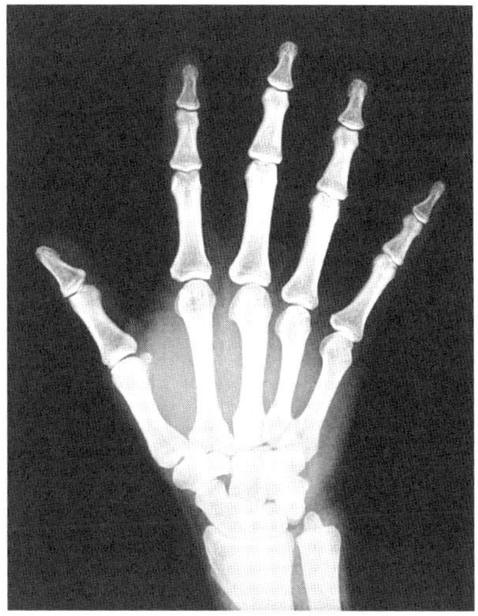

C5-8 Bones of the human hand

LEARNING EXPERIENCE
Fractions in the Human Hand

- Ask students to stretch out their hands and trace the outline, indicating the bone divisions of the fingers. Have them measure parts of the tracing and indicate the number of divisions found in the thumb and fingers. How can the hand be divided into fractions (for example, ⅓, ½,)? Measure each part of the hand, including each segment of the fingers. Have students compare with classmates their measurement findings. What relationships can be found between any two features of the hand? Does palm length give any indication of the length of the fingers? Is there a relationship between the thumb and little finger? Are any of the measurement ratios close to the golden ratio (i.e., approximately 1.62)?
- Vitruvius discovered that the length of the hand from the wrist to the top of the index finger is equal to the length of the head. Students can measure their hands and faces to try to confirm Vitruvius's findings. Have students compare their findings or create a graph.

LEARNING EXPERIENCE
Contour Drawings of the Human Hand

Contour drawing means drawing the edge of an object, including the folds and indentations, as a way of coordinating the hand and the eye, that is to say, what is seen with the eye and the movement of the drawing instrument. Contour drawings are line drawings that are developed shape-by-shape, detail-by-detail, to arrive at a completed form. As the eye follows the contours of the hand, the hand draws the line, trying to "feel" the edge of the object visually. (Refer to blind contour drawing, cross-contour drawing, and cross-hatching in the Glossary and Index of Key Visual Arts Terms.) In this activity, students will create contour and simplified drawings of their hand in order to study proportions and measurement.

▣ Suggest that your students look at one point of their hand, place their pencil on the paper, which will be the same place that their eyes are focusing on the hand, and as they move their eye slowly along the contour of the hand they should move their pencil on the paper at the same time. After creating several contour drawings of their hand in several positions, students can develop simplified drawings of the hand as geometric shapes, such as rectangles, lines, and parallelograms, or they can create a geometric drawing of their hand. Draw the hand with just straight lines or curved lines, or tubes and circles, or divide the shapes and spaces into coloured areas, or black, white, and grey areas.

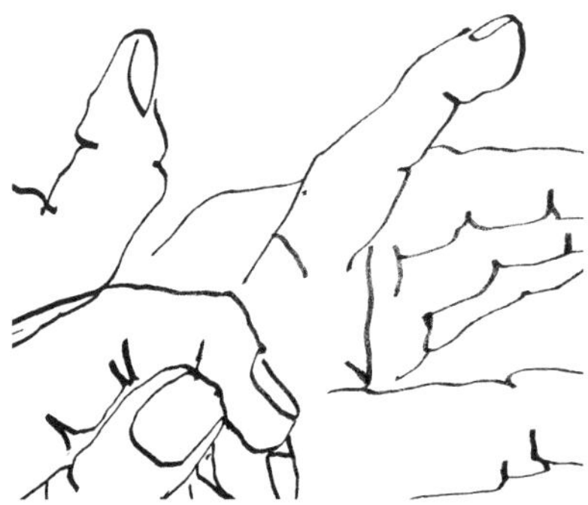

C5-9 Blind contour drawing of human hands

The Eye

Most human bodies, generally speaking, possess similar features and parts, but it is the differences in these features that make each individual unique. The eyes are one such example of a distinguishing human feature. It has been argued that by looking into someone's eyes, one can tell a great deal about the individual. The eye is a ball, or sphere, with skin flaps covering it on top and bottom. The eyeball is about 25 mm in diameter, and the white part is called the sclera. The eye is in constant motion when we are awake and asleep, especially when we are dreaming. We rarely really look at the eye and consider the sphere that gives us sight. There are many parts that make up the eye: the pupil, the iris, the cornea, the eyeball, the upper and lower eyelid and the eye duct.

LEARNING EXPERIENCE
Drawing the Human Eye

- Look closely at a person's eye. Notice how the upper lid folds over the eyeball and covers part of the iris. The eyelid is made up of soft tissue that takes the shape of the eyeball located underneath it. Notice the way in which the pupil reflects light in interesting ways, depending on the

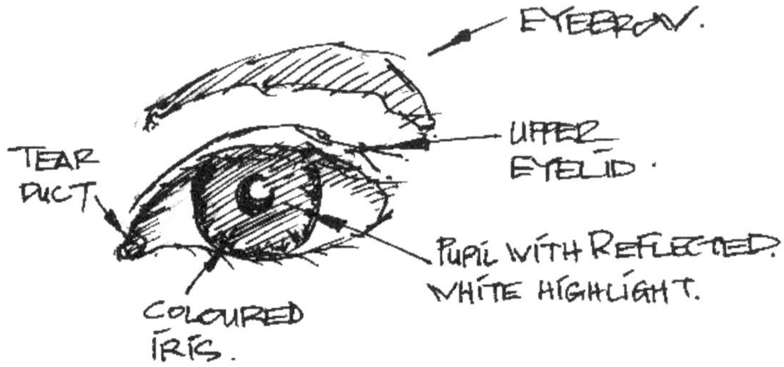

C5-10 Drawing of the human eye

light source and movement of the eye. What colours are reflected in the iris? Notice the light and shadow on the sphere of the eyeball.

- Sit opposite your model, no more than an arm's length away. Place the drawing paper on a drawing board so that it sits at a 45° angle or more. Have a variety of drawing pencils available from hard to soft (such as 6H, 4H, 2H, HB, 2B, 4B, and 6B) and an eraser (to increase the comfort level of the art student!) Have students draw an eye using the following procedure:

1. Draw the eyeball with a soft pencil (4B, 6B) looking for the shadows that exist around the edge of the eye and creating these shadows with the pencil.
2. Draw the pupil, with the white highlight of reflected light, using a variety of pencils for shading purposes. An eraser works well as a drawing tool to create the highlight, re-establishing the white of the page.
3. Draw the upper eyelid covering the eyeball. Notice if the eyelid covers

part of the iris. Does the shape formed by both eyelids look similar to an ellipse in geometry? Draw the lower eyelid. Does it meet at the eye duct or start a little above it?

4. The eyebrow is at the base of the forehead. Notice where the eyebrow starts and ends, and how thick the eyebrow is at the centre as opposed to the ends. Use the harder pencils (HB, 2H, 4H or 6H) to draw the hairs in detail.

- An alternative approach would be to measure the actual length and width of the eye and its composite parts, and then to draw these dimensions on grid paper. The student could then enlarge the image of the eye using larger grid paper.
- Extensions: Draw many different eyes, from different positions and angles. Choose eyes of people of different ages, genders, and ethnic origin. Really look at the individual parts that make up the eye. Include all the shadows, highlights, skin folds, and wrinkles!

The Foot

Leonardo da Vinci was so impressed with the design and function of the human foot that he considered the foot a masterpiece of engineering and a work of art. Our feet are marvelous instruments that we tend to take for granted. Athletes, especially runners, understand the importance of proper footcare, choosing their footwear carefully and keeping their feet healthy and regularly stretched. Humans have more bones in their feet than in any other part of their body. The bones of the foot include the five metatarsals, which form the instep, and fourteen phalanges, which form the toes (Marieb, 1987).

LEARNING EXPERIENCE

Ratio and Proportions in the Human Foot

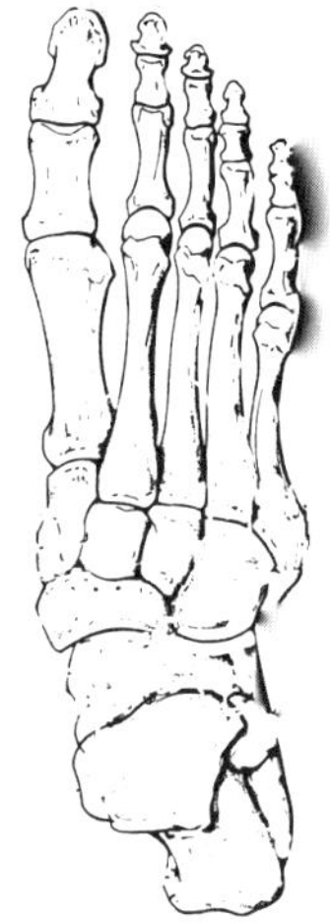

C5-11 Bones of the human foot

- Ask students to trace the outline of their bare feet and attempt to indicate the bone divisions of the toes. Have them measure parts of the tracing and indicate the number of divisions found in the toes.
- Have them measure each part of the foot, including the segments that make up the toes. Can students discover the golden ratio within the foot? (See page 162 for detailed instructions on how to determine if two measurements lead to a golden ratio of 1:1.61803 , or phi.)
- Have students create contour, simplified, and geometric drawings of their feet in order to study proportions and measurement.
- Vitruvius discovered that the foot is one-sixth the length of the body. Have students conduct research to determine if this is true. What are the variables that make this discovery debatable? Have students measure their foot. Using this measure as a ratio, ask them to determine how many feet tall they are. Have them compare their findings or create a graph.

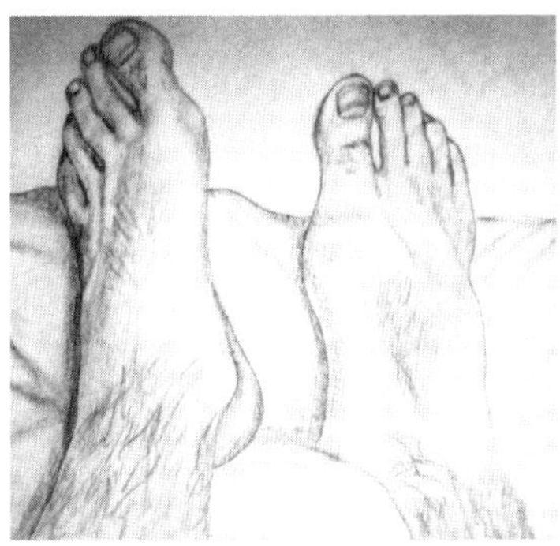

C5-12 Drawing of the human foot

The Ear

Located on either side of the head, the ears are ideally formed for hearing. The ears are larger than some would normally assume. From the front of the face, the top of the ear is approximately horizontal to the eyebrow and the bottom is horizontal to the lips. However, as with all other body parts, there are variations in size and shape. Many cultures alter the shape and length of the ears for aesthetic reasons, as well as pierce parts of the ear in order to carry and display artifacts or jewellery.

The parts of the ear that are most visible are the helix, which is the incurve rim at the top of the ear, the antihelix, the lobule (ear lobe), which is located at the very bottom part of the outer ear, and the crest of helix, which is the shape that follows inward and towards the face. The last visible part is the external auditory canal, which is the channel through which sound is led into the middle ear. The ear drum can sometimes be seen, or at least felt with a cotton swab. It is a thin layer of skin at the end of the canal.

LEARNING EXPERIENCE
Exploring the Human Ear

- Encourage students to really look closely at another person's ear. Have them create a contour drawing of the ear, starting with the outer shape of the ear. Draw the helix and the crest of the helix, and then from the helix, draw a line down to the ear lobe. Measure each part before and after it has been drawn. What geometric shapes do the helix and lobe resemble? Draw the ear just using geometric shapes.

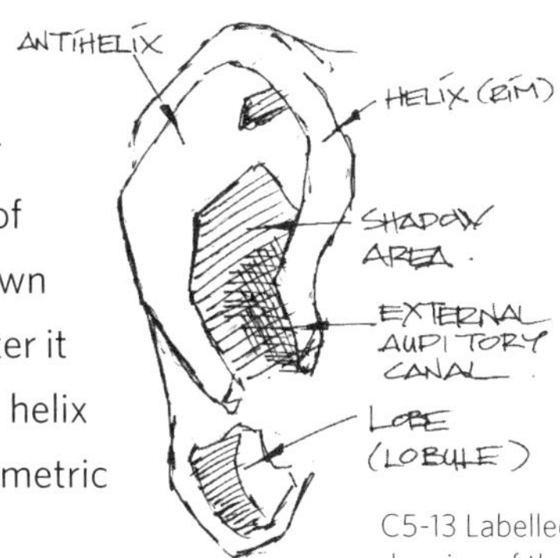

C5-13 Labelled drawing of the human ear

- Measure the ear from the top of the helix to the bottom of the ear lobe. Does the size of the person's ear have any other relationship to other parts of the head or to the hand? Measure the distance from the top of the helix to the antihelix, and then to the bottom of the ear lobe. What is the ratio between the two measurements? Gather data from a group of people. Create a graph or scatter plot. Does ear size change with the age of the individual?

Forensics and Human Identification

Although we all have similar parts (hands, fingers, palms, and feet), they are all different. *Forensics* is the art or study of argumentative discourse for a public forum or debate, suitable in courts of law. In other words, it is information or evidence that will "stand-up in court." This is why hospitals sometimes take babies' footprints and police take fingerprints of suspects for identification. These body parts have unique variations in lines, patterns, and shapes. Take a look at the lines in your palm. Do they run in the same direction, or do they intersect? What angles do they produce? Are the patterns in both palms the same? Students can draw hand palm patterns and collect fingerprints for study or observation.

LEARNING EXPERIENCE
Body Part Prints

- Measure and draw the open hand. Measure and draw a map of the palm lines on the drawing of the hand. Compare the lines for similarities and differences. Ink the fingers and palms to make prints. Consider what patterns emerge (repeated, regular, irregular, symmetrical, asymmetrical transformations). Students can use their body parts, such as fingertips, palms, feet, or heel of the hand or foot, as stamp prints to create patterns through repetition.

C5-14 Human fingerprints are used in forensics to identify criminals.

Proportion and Measurement

For centuries, artists have used the process of creating visual interpretations as a means of depicting their family and friends, capturing significant moments of social interaction, and more fully understanding themselves. Students should look closely at the proportions and details of people and faces. A good method for calculating proportions and measurements is the following: hold up a pencil and sight the tip of the pencil at the top edge of the object that you are measuring. Next, line up the end/bottom of the object being measured with the thumb on the pencil. This distance can now be used as a mental anchor for making other measurements in the drawing. Using this process, students can sight calculate or sight measure objects.

Vitruvius discovered that the distance from the chin to the base of the nose is one-third the length of the face, and is equal to the distance from the base of the nose to the bridge of the nose, and from the bridge of the nose to the top of the hairline (Verstockt, 1987). In other words, the adult human head can be divided into three parts: from the top of the head to the top of the eyebrows; from the eyebrows to the bottom of the nose; and from the bottom of the nose to the bottom of the chin. The human head can also be subdivided into two halves with a line drawn horizontally through the pupils of the eyes.

C5-15 Using a pencil to estimate and compare proportions in portrait drawing

The following are various methods of drawing portraits: detail-by-detail, using proportions, using a coordinate system, using grids (1:1 ratio and distortion grids), half portraits (reflections), silhouettes, caricatures, and stretch-and-bend distortions.

LEARNING EXPERIENCE

Portrait Drawing Using Proportions and Estimation

◻ Subjects can be used to start students drawing portraits and people. Drawing from live models, such as classmates or guest models, is best, but photos and other pictures can also be used. Self-portrait drawings can be done using a mirror or photograph.

This method of drawing uses proportions for an average adult head. Remember that it is the variations that give humans our individuality! Have a model, or use a photograph or a mirror for a self-portrait. The basic shape of a head is oval, approximately three units wide and three and a half units high. Create your own scale, for example, 6 mm = 1 unit. Using the following procedure:

1. Divide the oval in quarters across the width and length. The center horizontal line forms the eye line that runs through the middle of the pupils.
2. The eyes are approximately half a unit either side of the central vertical division, and each eye is approximately half a unit in length.

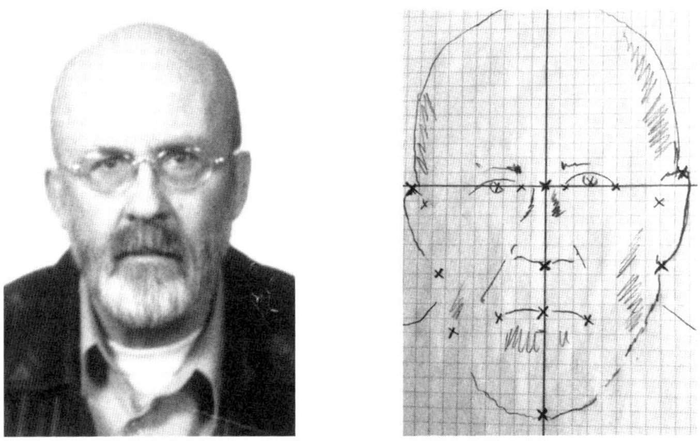

C5-16 Drawing the human face using a grid

3. The top of the nose is on the level of the eyes and ends one unit down.
4. The lips are half a unit down from the nose. The width of the lips lines up with the corner of the eyes.
5. The chin is half a unit down from the lips.
6. The eyebrows are approximately one quarter of a unit above the eyes.
7. The tops of the ears are in line with the eyebrows and end in line with the bottom of the nose.
8. The hairline is approximately half a unit from the top of the head.

This is an outline of the process of drawing a portrait. It is the artist who determines the unique details of the face being drawn.

LEARNING EXPERIENCE
Portrait Drawing Using Direct Measurement Transfer

▣ Adopting a direct measurement and transfer approach to portraiture simplifies the key elements of the face by using mathematical co-ordinates in order to quickly and accurately position key features of the face being drawn. This makes portraiture, arguably the most difficult realism drawing exercise, much less intimidating for the fledgling student of art. Since the measured distances of the head are transferred directly onto the drawing paper using a 1:1 ratio, no scaling or dilation of the model's face is necessary, again making the exercise easier for students. Jarvis (2010) recommends the following measurement/transfer approach:

> *Students pair up (assigned or self-selected pairs) and complete a measurement chart in which they actually measure distances: between pupils; from top of head to eye-line (pupils); from top of head to tip of nose; from top of head to lip-line; and, from top of head to bottom of chin. On large white Mayfair paper, students trace the contour of their partner's head,*

drawing a vertical line. They then map in their five measurements, adding an horizontal line at the eye-line and marking in the distance between the pupils on this line (i.e., divide the measurement by two and measure out on each side of the vertical line). Once this preliminary mapping is complete for each partner, students are now prepared to begin the portrait by carefully studying their partner's facial features and, beginning with the eyes, working from the inside out.... As a final step, colour is added using chalk pastels. Emphasize the working backwards and forwards with graphite (pencil), pastel, and eraser (all three of which are presented as equally important drawing tools) to slowly build up the portrait according to what they actually see when observing their partner's face, in terms of shapes and light/dark areas. When complete, these drawings are cut out and can be arranged in an overlapping collage of student [portraits]. (pp. 104–105)

Note that the four most common errors in portraiture are the following: (a) eyes too high on the face (usually in middle of forehead); (b) eyes too close together (not leaving a full eye width between them); (c) ears too small/short (like cartoon ears, instead of extending down to proper length); and (d) neck too thin (often drawn like a political cartoon with huge head on tiny neck).

C5-17 Instructor samples of direct measurement for human portraits
(colour plate xxiii)

LEARNING EXPERIENCE
Drawing and Scaling the Human Form Using Grids

For centuries, artists have used grids—a network of horizontal and vertical lines—to transfer an image into a smaller or larger form. For example, Albrecht Dürer (1471–1528) looked through an upright stationary wire grid and used a paper with an identical grid marked off. This helped him to observe and accurately draw his model proportionally.

In education, "copying" is most often viewed as negative behaviour, and rightly so when speaking of plagiarism and the reproducing of another's work without proper credit being given. In visual art, however, copying is a legitimate stage in an art student's development and has been done for thousands of years as a training method for artists working as apprentices under master artists. While travelling Europe, one of the authors witnessed, on more than one occasion, university art students setting up in major national galleries to copy the works of the masters as part of their visual art curriculum and training. Students can learn a great deal through completing simple to complex copying exercises. These works are not, of course, treated as original works done by the artist, but rather function as technique studies that help students of art to begin to develop their own particular style involving combinations of different techniques.

- When asking students to copy from a picture, photograph, or artist's reproduction of drawings or paintings, have specific objectives in mind, such as an emphasis on drawing and shading techniques, or understanding the mathematical concepts involved in using the grid (proportion, enlargement, magnification, distortion).
- Drawings can be enlarged or reduced (grid up or grid down) or distorted using a distortion grid. One trick for helping students use a grid is to cre-

C5-18 Leonardo da Vinci used grids to enlarge drawings for fresco paintings.

ate a transparency of a grid and place this over the picture to be drawn (Naested, 2002, p. 121). Have students use grids to enlarge famous paintings (or selected sections) of human portraits or figures.

Beauty and the Human Form

Historically, numbers, shapes, mathematical formulas, relationships, and proportions have been used in an attempt to describe what is ideal and beautiful. Perhaps most famous among these ideas is the concept of the golden ratio, or phi, (Greek letter) of approximately 1:1.6. Marquardt, a maxillofacial or "plastic" surgeon, developed a "mask" or template for human beauty based on phi and the regular pentagon. It is claimed that he conducted cross-cultural surveys and found that all groups surveyed had the same perceptions of facial beauty (Phi-Nest, n.d.). Some artists/cultures throughout history have maintained that the ideal human form is seven heads tall, some claim it is eight, and others contend it is ten heads high. While there does not appear to be historical consensus on this point, it is certainly true that the topic of ideal human proportions in the body and face have been discussed and depicted by philosophers and artists in many cultures throughout human history. Take, for example, the ancient Greeks, who used the golden proportion to create their female Caryatid sculptures that formed columns within the porch of the Erechtheion structure on the Acropolis mount in Athens.

Symmetry of the Body and Face

The word symmetry can mean well-proportioned, well-balanced, or having the concordance of several parts that are integrated into a whole. There are different kinds of symmetry. Symmetry and balance are linked to bilateral symmetry, the symmetry of left and right, which is a prominent feature of the higher animals, especially the human body (Weyl, 1969). There are, however, subtle differences or exceptions to the "mirror image" of human symmetry: the heart is not central, but is located to one side; there is a predomination of right-handedness, which results in uneven muscle distribution; and the two sides of the face are not identical, but only approximate. According to Stewart and Golubitsky (1992), this is why reflection of a person's face and body in a mirror can look quite different to the viewer. "But the overall form is very close to one that has perfect symmetry" (Stewart, 1995, p. 74). Although symmetry is considered beautiful in most cultures, the idealized human figure with both sides of the body being exactly the same does not really happen.

LEARNING EXPERIENCE
Portrait Reflections

- Collect pictures of portraits, photocopy the originals, and then cut each portrait in half. Have students attempt to draw the exact mirror image, or reflection, of one side of the portrait. Then have students compare their mirror image drawings with the original portrait, noting any differences between the two images. Note: Since all human faces (and bodies) are not perfectly symmetrical, this exercise often leads to some interesting observations: famous actors/celebrities, or familiar faces for students, often do not quite "look right" when the reflection drawing is completed with a high degree of accuracy since in reality, the face students are used to seeing is not perfectly symmetrical. Facilitate a class discussion on this point. Encourage students to use a variety of pencils (soft such as 6B, 4B, and 2B, or hard such as 2H, 4H, and 6H) to achieve the light and dark tones of the face.

LEARNING EXPERIENCE
Silhouette Portraits

A silhouette is an outline of something, such as a human profile, filled in with a solid colour. Using a light source such as an overhead projector, students can trace each other's silhouette on paper mounted on the wall. This learning experience can be connected to mathematical concepts such as projection and dilatation. Projection is the process or technique of reproducing a spatial object upon a plane or curved surface by projecting its points. Dilatation is the transformation of a figure, re-creating a new form that is similar in shape to the original, but either larger or smaller that the original. The area of the silhouette can be calculated by placing or drawing the silhouette image on a grid or graph paper and counting the full and partial squares. Students can experiment with whether size varies depending on the distance between the projector and paper.

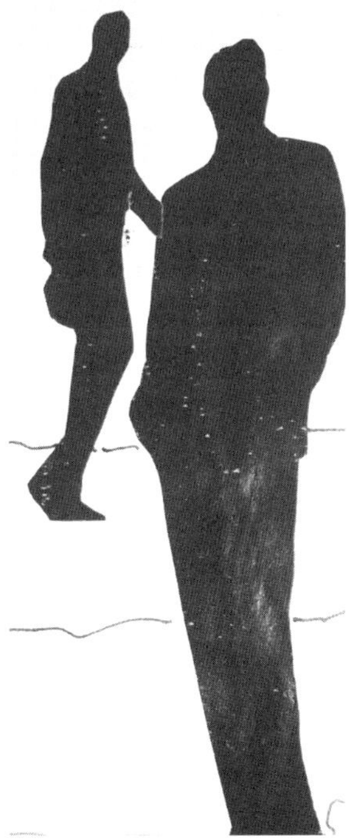

C5-19 Silhouettes of the human form

- Students can use their profile as a place to mount pictures and images that represent who they are, what their strengths are and what they enjoy doing. Two images, one in black and one in white, could be images of their positive strengths and things they would like to work on.

LEARNING EXPERIENCE
Gesture Drawing

Gesture drawings are quick drawings produced by artists and art students for several reasons. One reason is to loosen up the hand, become familiar with the drawing instrument, and prepare for a longer drawing session with a human model. Another is to draw the figure as quickly as possible, getting the essential shapes and proportions before the figure moves on (as in the drawing of a figure involved in a dance or sport). In gesture drawing, you draw not what the figure looks like, but instead what it is doing. This is also an exercise in estimating. An estimate is a rough decision or judgment on the number, amount, value, or size of an object.

In this exercise, students will observe and draw a figure in proportion using the gesture drawing technique, thereby capturing the figure in motion, and coming to understand the proportions of a figure.

- Demonstrate the gesture drawing technique for students on a board or flip chart while looking at a model. The model for the gesture drawing could be a classmate, a classmate in costume, the instructor, or a guest model (as an alternative to using students or guests as models, students can use photographs of moving figures, art mannequins, a large doll, or even a stuffed animal such as a bear or monkey). The materials students can use are varied. If students are unfamiliar with the drawing instruments or surfaces they will be using, it is important to demonstrate the activity and give students some time to manipulate the instruments.

- Position students so that they are facing the model, with drawing boards held at a 45° angle. The model should be elevated and positioned so that all students can view his or her full height. Time the poses, from longer (1 to 2 minutes) to shorter periods. Have the model move into positions from low to high, from front to back to side (rotation). Encourage students to ghost draw first for 10 to 15 seconds, planning the size and shape of the figure on the page by moving the arm and drawing tool around the board, but hovering above the surface of the paper without actually making marks. Once the actual drawing begins, encourage students to focus on the skeletal bone structure, quickly pencilling in the location of the spine, arms, legs, and head without showing any details or paying attention to outside contour lines. Students can draw and overlap many poses on one piece of drawing paper.

C5-20 Gesture and contour drawing

- Students could also use the end of their pencils to sight measure various parts. For example, they could estimate the size of the head and use this as a ratio to discover the height of the model or the width of the shoulders.
- Discuss the drawings after the session. Look for proportion and angle, as well as for balance on the supporting body parts, such as the legs (i.e., where is the weight of the body, often shown by darker lines?). In order to facilitate the sharing of ideas, stu-

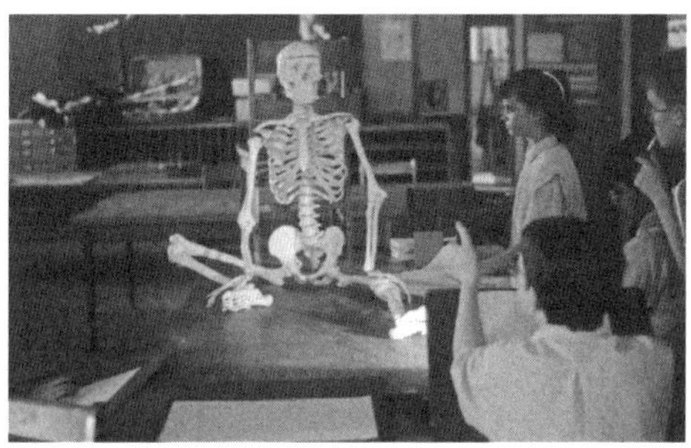

C5-21 Observing the human skeleton in drawing class (colour plate xxiii)

dents should be permitted to walk quickly around the studio/classroom to see the various interpretations of the same set of poses by their fellow learners.

LEARNING EXPERIENCES

Blind Contour and Mass Drawings of the Body

- Blind contour drawing (also called continuous line drawing) involves keeping the eye on the form while the pencil delineates the form on paper. Blind contour drawings are sometimes done using translucent wax as a drawing instrument (using wax prevents students from viewing their drawing while it's in progress). The drawings are then painted over with watery paint to bring out the form drawn in wax.
- Mass drawings are meant to render the solidity or mass of the subject using areas of tone or colour, without emphasizing lines or edges. The model is set up with a spotlight to create contrasting areas of light and shadow. Students can either work with charcoal, black conté, or pastel on light coloured paper; or with white chalk, light-coloured conté, or pastel on dark paper.

LEARNING EXPERIENCE

The Human Body Made Geometric

- Geometric drawings are drawings only using geometric shapes to create the image. After studying the human form, students can recreate the form with geometric shapes such as circles, triangles, rectangles, and squares made from coloured paper or cloth. These figures can be made into robots or strange creatures. Instead of being glued onto paper, these shapes can be strung together to create a mobile.

LEARNING EXPERIENCE
Portraits in Caricature and with Distortion

Creating cartoons is often a favourite exercise for the young artist. Once students are aware of the detail of the human portrait, you may wish to introduce the topic of caricature. A definition for caricature is the deliberate distortion of a figure through depicting a person with exaggerated features. In order to exaggerate features, the artist must be familiar with the features of the person they are caricaturizing. This can lead to creating cartoons and flip books that illustrate motion, transformation, and translation.

- Have students research different forms of caricature drawing, and then select a photo and create a caricature of the image using the techniques they have learned.
- After drawing the human face from observation, students can take one of their drawings and gradually distort the features by stretching parts in several phases, using translations or metamorphoses to create an entirely different image. Software such as Adobe Photoshop can also be used to alter a portrait image with special effects.

C5-22 Stylized group portrait using Adobe Photoshop effects

Sculpture and the Human Figure

The following are suggestions for sculptural projects using various media, including wire, papier mâché, clay, fabric for soft sculpture, shapes, and people boxes.

LEARNING EXPERIENCE
Wire Sculpture and Action Figures

▣ Refer to page 143 for general information about creating a wire sculpture. The materials needed are pre-cut 300 mm lengths of thin wire, one for each student. Rulers are also useful. The base for the wire sculpture can be made from scrap plywood pieces, rocks, clay, or plaster. The procedure is as follows:

1. Fold the wire in half, the bend representing the top of the head. From the top of the head, measure down 45 mm and bend the two ends of the wire back up.
2. Twist the ends around to make a neck approximately 8 mm from the top of the head.
3. Twist the wire together where the legs would meet.
4. Place the figure in an "action" pose (from one of the gesture drawings created previously), or in the position of a live model. The form can be stapled to the base, stuck into clay or playdough, or may be hot-glued to a rock or other solid material.
5. The wire sculpture can become an armature for the application of papier mâché.

C5-23 Wire action figure sculptures (colour plate xxiii)

Full-Sized Figures

Greek and Roman art generally showed humanity as perfect and ideal. More recent artists often depict humans as they really are: joyful or depressed, young or old, rich or poor, healthy or otherwise. In the 1950s, artist George Segal developed full-figure forms cast in plaster of actual people, and displayed the forms in their simulated "realistic" environment, such as in an elevator or at a lunch counter.

Most students will be thrilled to be allowed to work big! You may not wish to have your students make full-figure moulds in plaster (as Segal did) in your classroom, but there are many other alternatives to working large. If you are worried you don't have enough storage space during construction, try hanging the sculptures from the ceiling. Following are a few options for producing large sculptures. All require students to measure and to calculate.

LEARNING EXPERIENCE
Papier-Mâché and Fabric Soft Sculptures

C5-24 Papier mâché life-size sculpture of the victim in Goya's *The Third of May, 1808*. (colour plate xxiv) To view Goya's original, use Google Art Project. See page 291 for instructions.

- Have students construct papier-mâché sculptures made over wood, cardboard boxes, balloons, or chicken wire.
- Soft sculpture is a process of sculpting in fabric by stitching material together, stuffing the stitched fabric with material to make a three-dimensional piece that has height, width, and depth. Have students construct fabric soft sculptures made from cast-off clothing, scraps, yarn, nylons, string, needles and thread, a glue gun, and something to stuff the fabric with, such as newspaper strips or shredded paper. Students will need

to estimate the amount of material needed to make the soft sculpture and cut the fabric according to a pre-drawn pattern.

LEARNING EXPERIENCE

Caricature Faces

A caricature of the human face—a deliberate distortion of features, exaggerating or overstating certain qualities or features, or minimizing or understating others—can be an imaginative and unique learning experience.

- Cut 25 mm of nylon hose, stuff it with other nylons or other stuffing material, and fold and stitch the ends to the back, creating a ball shape. Determine through measurement where the nose should be located. Stitch along the side of the nose from the bridge to the end of the nose in a triangle. Stitch the bottom of the nose. Estimate where the eyes should be and sew beads or other round objects to represent the eyeballs. To form eyelids, fold some of the soft fabric beside the bead and stitch at the corners to create the eye duct and the opposite corner of the eyelid. Decide where the lips should be, and fold some of the soft fabric and then stitch the lips. Measure where the ears should be, then fold and stitch. Add other details such as cheeks and dimples. Glue or stitch on hair or moustache made from wool or string. Add earrings, jewellery, lace, hats, or ribbons. Consider using make-up, eyeglasses, or sunglasses to make the portrait even more interesting.
- A body can be made for this face using the same technique. An alternative is to use clothing that is an appropriate size for the head. A shirt and pants can be attached together and stuffed with newspaper, cloth, or other filler to create a body for the head.

LEARNING EXPERIENCE
Creating People Boxes

- After studying the human form, have students construct human sculptures out of boxes that are of similar proportions to the human form. These representations may become more like robots or strange creatures. The experience will encourage students to compare shapes, sizes, and proportions. Decorate, elaborate, and create detail. For example, students can paint the boxes or add found objects, such as coat hanger wire or yarn.

C5-25 People box sculptures with found objects (colour plate xxiv)

Disproportional Human Figures

Most of this chapter has dealt with human proportions and we have noted that often these proportions possess a certain amount of consistency, thus enabling art students to increase their skills in accurately depicting the human form or face. However, in this next section we discuss learning experiences that allow for more imaginative renderings of the human form, wherein students can explore disproportional human figures often found within the realm of myth and legend.

LEARNING EXPERIENCE
The Human Body and Metamorphosis

- After drawing the human body, students could take one of their drawings and, in stages through several drawings, recreate a very different image through metamorphosis using distortion and translation (e.g., a young boy changed into an old man, baby, monster, or animal).

LEARNING EXPERIENCE

Drawing a Sasquatch

The Sasquatch* can be thought of as an aggressive monster, a mythical creature, or a primate that walks on two legs and lives in northwestern North America. It has often been referred to as "Big Foot" and has been studied under "Crypto-bipedal-Primatology." According to Trapp (n.d.), this animal was seen by Aboriginal people in North America before the arrival of Europeans to the continent. As well as reported sightings, screams in the night (unlike any other animal's sounds) have been recorded, and footprints much bigger than a human's have reportedly been found. These alleged footprints also indicate the creature has a much longer stride than humans. The size of the Sasquatch has been claimed to be much larger than any human, possibly seven or eight feet tall, and the creature is often reported as being covered in fur like a oversized bear.

C5-26 Sasquatch drawing

* For more information on the Sasquatch, contact the Sasquatch Information Society headquarters in Seattle, Washington (http://www.bigfootinfo.org), or refer to books such as *The Field Guide to Bigfoot, Yeti and Other Mystery Primates Worldwide* (Coleman & Huyghe, 1999) to be found in one's local library or on the Internet.

In this activity, students will develop their interpretation of the size of a Sasquatch through mathematical problem solving.

- Make several footprints of various sizes that are to represent the footprints of a Sasquatch. These can be cut out from cardboard or moulded in clay or plaster. Give a pair or group of students a footprint. The problem is to determine the size of the Sasquatch from the footprint, including the height and head size. Students can make a scale drawing of their creature, or a full-size drawing using large paper. Other alternatives would be to create a soft sculpture, a clay sculpture, a playdough model of their creature, and to construct an environmental diorama for it.
- Alternative project: To begin, have students measure how many hands tall they are. (For example, a student could be eight hands tall). Give students a drawing or mould of a "Sasquatch" paw or claw. Have them measure this paw or claw and multiply the measurement by the number of hands high they have measured their own height to be. Students can then draw their Sasquatch using the ratio. How large or small would their creature be given different starting measurements? Students can produce scale drawings, or life-size images. They can complete their creature by painting it or by covering it with a collage technique using various materials, papers, yarns, or fabrics. You can have them create a clay, Plasticine or playdough model.

LEARNING EXPERIENCE
Leprechauns and Proportion

Fairies, faery people, elves, and leprechauns come from myths and legends of Ireland. They are believed to live just below the surface of the ground, and the prime time for them to be seen is around Halloween and May Eve. The leprechaun looks like a small human, about two feet tall, with a large head and pointed ears, and is often dressed in green with a hat. His toes turn up. It is believed by some that leprechauns have hidden pots of gold.

- A normal child's body is four or five heads tall. Considering that the mythical leprechaun's head is larger than a child's head, decide on the proportions necessary to create a Leprechaun. How large is his head compared to his feet? Compared to his body? Compare the head to his arms, legs, and hands. Create a drawing, painting, or a sculpture of clay, playdough or papier mâché.
- Similar learning experiences can be had with other creatures, either larger or smaller, such as giants, goblins, witches, wizards, etc.
- The study of a culture's myths and legends can suggest many creatures of invention. This would be an ideal connection to social studies and language arts. Students could create their own creatures, myths, and legends.

Summary

In Chapter 5, we explored the connections between mathematics and visual art found within the human form. This connection appeared to begin with early humans as they developed measurement systems based on the human body and body parts. The understanding of the human form and its related proportions has assisted and provided inspiration for the artist since the dawn of human civilization. We have also noted that several key mathematical concepts are connected to the creation of art images relating to the human form, such as ratio, proportion, balance, symmetry, measurement, transformation, and disproportion. A variety of learning experiences, such as realism drawing, portraiture, figure drawing, and abstract sculpture have been suggested for teachers and students to explore as they study the human form and its many complex patterns.

6 Mathematics and Art in the Designed Environment

Since man lives in a series of environments that are partially or totally man-made, it is fair to ask: Where does environmental form come from? The origin of man-generated physical form . . . lies in the interaction of human needs, attitudes, and values with natural laws and materials. Both man-made and natural objects can be read as a record of the forces that introduced them. (Preble, 1973, p. 356)

Introduction

There are few environments on earth that have not been altered at least somewhat by humans. According to Bartuska and Young (1994), "The world is now the human created environment—a built environment" (p. ix). This chapter introduces the rich and complex connections between mathematics and designed environments. These connections began with early humans as they developed measurement systems based on the human body and body parts. Their measurement systems assisted in the designing of spaces and objects to meet physical, social, emotional, and aesthetic needs. These systems helped to describe lengths and quantities by using body parts such as fingers and hands, and body movements to measure pacing when considering size, scale, function, and materials.

C6-1 Westminster Abbey façade in London, UK
(colour plate xxiv)

The discipline of architecture is the art, mathematics, and science of planning and building structures—where form has met function over thousands of years of human history. Like sculpture, a structure is a three-dimensional form, utilizing space, mass, texture, line, light, and colour—the elements that engage the emotions and senses. Learning experiences for mathematics and visual arts, as well as key concepts and images for process and product, appear throughout the chapter.

Humans build environments that satisfy human needs, wants and values (Bartuska & Young, 1994, p. 8). Depending on the level of needs according to Maslow's hierarchy of needs theory (1970), the built environment may simply fulfill needs on the subsistence level, such as the need for shelter, or may also address much higher level needs, such as self-actualization (e.g., artistic expression).

Architecture evolved to meet humans' need for shelter, while attention to design met the human need for an aesthetically pleasing environment that complemented the design site and was suited to the climate (Preble, 1973). Structures are also designed for various purposes. The "vernacular" structures served the "ordinary" people in their daily lives. The "high-style" belonged "to the elite, to governments, religious cults and institutions" (Bartuska & Young, 1994, p. 13). Architects consider not only solid shapes and surface decoration, but also the spaces in and around the shapes. They consider the materials to be used and often oversee the construction.

Architecture and Geometry

"In architecture, the rules of both plane and solid geometry apply to three-dimensional shapes and spaces. When height or thickness is added, plane geometrics on paper become rooms, buildings and open spaces" (Blackwell, 1984, p. 5). These open spaces that are created are not solids but voids, or volumes, for habitation. Geometric shapes are combined and used in architecture; the shape of the buildings and roofs, such as pyramids, cones, prisms, triangles, rectangles, hexagons, pentagons, and circles. Use of the various combinations of geometric forms changed over time, cultures, and climates, to create "architectural styles."

The following are some of the most prominent geometric shapes found in architecture. It may help to understand how geometry brings a sense of "compositional cohesiveness" where the elements of the work have a visual sense of belonging. "By revealing some of the geometry, systems, and proportions it is possible to understand better the intent and reasoning of a number of designers and architects" (Elam, 2001, p. 43).

Circle: The circle has often been considered of symbolic or ritual importance, such as in the Stonehenge monument, conference round tables, monuments, music and drumming circles, circular drama and dance performances, stadiums and amphitheaters (e.g., the Coliseum in Rome (AD 72–80) and the circle plan of the Temple of Sibyl, Tivoli (early first century BC).

According to Blackwell (1984), "When uniform pressure is exerted on a circle, either from the inside or from the outside, the circle is the strongest of shapes" (p. 97). The circle has been incorporated into architectural features, such as the rose windows in Renaissance designed cathedrals, or even in the fictional door of the hole-dwelling Hobbit in Tolkien's classic adventure. The symbolic importance of circles will be presented in a later chapter.

C6-2 Circle design of Chi-Rho monogram of early Christianity

C6-3 Circle design of outdoor firepit

C6-4 Circular design of the London Eye Ferris wheel in London, UK (colour plate xxiv)

Triangle: The secret of the triangle as a shape used in buildings and roofs is that each side put together helps to strengthen the others. The equilateral triangle is a versatile and frequently used shape for shelter throughout the world, from simple tents and bivouacs to much more elaborate structures. Roofing structures built with triangles protect humans from the natural elements of sun, wind, rain, and snow. The Egyptians used right triangles called the "Egyptian Triangle" in constructing the pyramid. The Parthenon on the Acropolis in Athens features triangular pediments.

Newton's Mathematical Bridge, as it is traditionally called (although Newton, a Cambridge professor, apparently had nothing to do with its construction),

C6-5 Triangle design in roof trusses provides structural support (colour plate xxv)

C6-6 Triangle design in wooden supports for Newton's mathematical bridge in Cambridge, UK (colour plate xxv)

features many triangles in the wooden support structure, located next to Queen's College at Cambridge University, UK.

Rectangle: The golden section—that fascinating number relationship that has intrigued mathematicians since antiquity—can be derived by dividing a straight line into two unequal segments (*a* and *b*, where *a* is smaller than *b*), such that the ratio of the smaller section to the larger section (*a:b*) is equal to the ratio of the larger section to the entire line segment (*b*:(*a* + *b*)). When these two ratios are set in proportion,

$$\frac{a}{b} = \frac{b}{(a+b)}$$

and we also introduce the idea of $a = 1$ unit and $b = x$ (some unknown length), then we now have

$$\frac{1}{x} = \frac{x}{(x+1)}$$

and so can use the Quadratic Formula

$$x = \frac{-b \pm \sqrt{b^2 - 4ac}}{2a}$$

to solve the single equation for the unknown variable. In this case, *x* simplifies to

$$\frac{1 \pm \sqrt{5}}{2}$$

but we consider only the positive solution for sake of context (i.e., the negative length is meaningless in this model), and so "*x*" is approximately equal to 1.6180339. This means that in order to create a golden section relationship, one need only multiple any given line length by "phi" or 1.6180339. In rectangles, the golden ratio of length to width is considered pleasing to the eye. To analyze any rectangle to see if it is "golden," one need only divide the length by the width and see if the quotient is near phi (1.6180339) (Jarvis, 2002, 2007). Marcus Vitruvius Pollio (31 BC–AD 14) applauded the aesthetics of the divine proportion in his book *De Architectura*. Many buildings are based on this golden rectangle, not only for their outer shapes, but also for the spaces within rooms, corridors, doors, and windows. The Parthenon in ancient Greece featured this design element in multiple facets of the floor and building plans. The west façade of Notre-Dame Cathedral in Paris (1163–1250) has golden rectangles in its design, as well as circles, arches, and triangles ("flying buttress" supports). Further examples of the golden section in design include Frank Lloyd Wright's Robie House in Chicago (1909), and Walter Gropius's Shop Block in Dessau, Germany (1925).

Square: A square is a flat shape with four straight and equal sides and four interior right angles. Similar to a circle being categorized as a special case ellipse (i.e., where the two foci are coincident), a square is actually, by definition, a special case rectangle where all sides are of equal length. The walls of regular prisms, the surfaces of cubes, and the bases of four-sided pyramids are all squares. The "square of 4 is 4 x 4 = 16. Similarly the cube is what you get when you multiply it by itself twice; for example, the cube of 4 is 4 x 4 x 4 = 64" (Stewart, 1995, p. 5). Santa Sophia in Istanbul (AD 537) has a square central space that is covered by a dome. Square grids are also used in the organization of designed landscapes, such as city squares (e.g., the Red Square in Moscow, St. Peter's Square in Rome, Aztec cities in Mexico). Upon their arrival in North and South America, many European surveyors planned square townships and one-square-mile farms.

Pentagon: A pentagon is a five-sided shape. A ring of ten regular pentagons can be used as the main structure of a building or roof design. A well-known US

C6-7 Rectangular designs in the complex façade of Notre Dame Cathedral in Paris, France (colour plate xxv)

C6-8 Pentagon design of the Pentagon in Washington, DC (colour plate xxv)

government building constructed in 1941–43—The Pentagon—is named after its characteristic shape. One of the world's largest office buildings, this structure houses the US Department of Defense and was designed by George Bergstrom.

Hexagon: The equilateral triangle forms the basis of a regular hexagon (i.e., all sides are equal in length, and the regular hexagon contains six equilateral triangles). Frank Lloyd Wright used hexagonal themes for his design of buildings and furniture (Blackwell, 1984).

C6-9 Hexagon design in table umbrella (colour plate xxvi)

Octagon: Orson Squires Fowler thought that the perfect form was a circle, but decided it would be easier to build an octagonal house on his New York property. The octagonal design became popular across North America in the mid-1800s (Ross, 1996). The regular octagon has eight equal sides and eight equal angles. Many other names exist for shapes of different side numbers. For example, the hexadecagon (16 sides), the chiliagon (1,000 sides), and the myriagon (10,000 sides).

C6-10 Octagon design in picnic table (colour plate xxvi)

Dome, sphere, and semi-circle: Semi- or half-circles occur in the design of Inuit snow structures (commonly referred to as "Igloos") of northern Canada. It was also a traditional shape of the huts fashioned by Zulu and Swazi tribes of Africa. Of military origin, the Quonset hut is

C6-11 Hexadecagon design in outdoor pool (colour plate xxvi)

C6-12 Dome design of Inuit igloo constructed with snow and ice (colour plate xxvi)

C6-13 Dome design of the Dome of the Rock Islamic Mosque on Temple Mount in Jerusalem (colour plate xxvii)

a self-supporting structure in a curved or arched shape. In formal buildings, the dome appeared in amphitheatres and arches in the Hagia Sophia in Istanbul (AD 532–537) and the Pantheon in Rome (448–432 BC).

Spirals and irregular curves: A spiral is a curved line that moves away from, yet also around a fixed point. Spirals appear often in natural forms, such as the growth patterns of leaves, sunflowers, and the chambered nautilus. Architectural examples include the Colosseum in Rome (72–80 AD), the spiral tower of the Mosque of Muta Wakkil (AD 848–52) in Smarra, Iraq (*Tower of Babel* painted by Bruegel), the Cathedral of St. Basil (1554–60), Moscow, and the Lansdowne Crescent in Bath, England (1789-93). The Taj Mahal (1630–48 AD), located in Agra, India, has an elegant dome and projecting cornice. From 1950–55, Le Corbusier designed and built the Notre-Dame-du-Haut in Ronchamp, France, with curved rooms and roof. Another notable spiral space is found in New York's Guggenheim Museum, designed by architect Frank Lloyd Wright in 1959.

Pyramids and cones: Ancient Mesopotamians built pyramid-style structures in the form of receding tiers upon rectangular, oval, or square platforms called "Ziggurats" (end of the third to the sixth century BC). Early Egyptians developed the pyramid construction. The great pyramids of Giza (4,600 years old) are considered one of the seven wonders of the ancient world.

The shape of pyramids and cones are proportional and aesthetically pleasing. Their volume is found by multiplying the area of the base by one third of the height, "a relationship proved by the Greek geometer and astronomer Eudoxus of Cnidus (408–353 BC)" (Blackwell, 1984, p. 141). This geometric relationship can be easily modelled and experienced by constructing a cube and a square-based pyramid (i.e., with a base and height identical to the cube), then filling the latter with sand or rice and pouring it into the former figure. There should be exactly enough sand/rice in three filled and levelled-off pyramids (positioned upside-down, of course, to carry the sand/rice!) to completely fill the cube. The Sioux teepee (or tipi) is a circular-based pyramid, or cone, arranged with eighteen supporting poles.

Architect Ieoh Ming Pei designed the glass pyramid at the entrance building of the Louvre in Paris, France. The Louvre Pyramid is a large glass and metal pyramid surrounded by three smaller pyramids in the main courtyard (Cour Napoleon) of the Louvre Palace in Paris. The large pyramid serves as the main entrance to the Louvre Museum. Completed in 1989, it has become a landmark of the city of Paris. The construction of the pyramid triggered considerable controversy because many people felt that the futuristic edifice looked quite out of

C6-14 Egyptian pyramids (colour plate xxvii)

C6-15 Glass pyramid of the Louvre Museum in Paris, France (colour plate xxvii)

place in front of the Louvre Museum's classical architecture. Others lauded the juxtaposition of contrasting architectural styles as a successful merger of the old and the new, the classical and the ultra-modern (Metcalfe, n.d.).

LEARNING EXPERIENCE
Architecture Field Trip and Space Design

- Take your students on a field trip in order to identify and to draw various architectural features that include geometric shapes. Look for features such as pillars, archways, roofs, windows, and doors.
- Study architecture from various cultures and periods to determine what geometric shapes were used in the design of structures or public/private spaces.
- Use geometric shapes to design a building for a specific purpose.

European Influences on Architecture

Vitruvius is said to have established three main requirements for architecture: durability, convenience, and beauty. The most elusive and subjective quality is arguably that of beauty (Martienssen, 1976). Style, fashion, and cultural norms form important elements in what is considered by a person or population to be aesthetically beautiful. According to John Ruskin (1819–1900), architectural style is adapted to the situation and climate, and has strong affinities with "the prevailing turn of mind by which the nation who first employed it is distinguished" (Bradley, 1971, p. 13). In North America, early architectural influences comprised mainly ancient Greece and Rome, and the Gothic and Renaissance periods, rather than the eastern styles of Japan, China, and India, or the Aboriginal styles of South and North America. As emerging nations, Australia, Canada, and the United States relied heavily on what was considered to be "great" and "beautiful" by European standards in both architecture and visual art. The following is not meant as an exhaustive list, but rather provides a brief sampling of investigations that students may wish to undertake when studying the architecture of various countries and cultures.

Egyptian (2000–100 BC): With their knowledge of astronomy and mathematics, the Egyptians used geometry and repetition as symbols of stability and permanence. Some give credit to the Egyptians as the first to use mathematics in their art and architecture. For example, they created monuments, pyramids, and giant sphinxes.

Classical Greek (1100–350 BC) and Roman (150 BC–AD 1000): The golden ratio as a numeric value is called phi, named for the Greek sculptor Phidias. In Greek architecture, the spaces between the columns often formed golden rectangles. The ancient Greeks believed the golden rectangle to be a pleasing shape due to the ratio of the longer side to the shorter. The Parthenon, built in Athens (fifth century BC), is a renowned historic site featuring many instances of the golden ratio in its design.

There are three prominent styles, or orders, that appear in Greek architecture: Doric, Ionic, and Corinthian (see Figure C6-16). The Parthenon is most famous for its Doric columns. Doric columns have no base, but rise directly from the flat pavement (the “stylobate”). The shafts usually have twenty shallow flutes with sharp ridges between and intricate patterns of fluting. Some are carved with a leaf pattern. Ionic columns are taller than Doric and have flutes that are lines carved into the columns from top to bottom. The frieze is plain, the bases appear to be stacked onto rings, and on the capitals are scrolls. The Corinthian column is considered to be the most ornate and elegant. One of the earliest examples of the Corinthian column is the Temple of Apollo in Bassai. The basic shape is that of an inverted bell. This order has an advantage over all the other orders because its decorative features can be viewed equally well from any side. The faces of the capital between the volutes had relief decoration of spirals and a palmette. A palmette is a decorative motif based on the fan-shaped leaf of a palm tree. Corinthian capitals are more commonly employed towards the end of the fourth century AD and in the Hellenistic Period (180–31 BC).

The Greek outdoor theatre is known for its functionality and aesthetically pleasing design. The structure of the theatre consisted of three main parts: the orchestra, the theatron (or auditorium), and the skene (or dressing room). The orchestra was an important place for the chorus forming the centre around all the other parts of the structure. This was the main stage for the chorus and the actors situated in front of the spectators. The auditorium was built with rows of seats for the spectators that were stacked and offset, thereby allowing all to see and hear what was happening on stage.

The Romans appropriated and often further developed many Greek architectural designs, elements, and ideas. Roman architects (or what we now consider engineers) are, however, credited with the invention of the arch, the vault, and the dome.

Renaissance and Baroque (1420–1860): Filippo Brunelleschi (1377–1446), Leone Battista Alberti (1404–1472), Leonardo da Vinci (1452–1519) and Michelangelo Buonarroti (1475–1564) were Italian artists and designers who reinterpreted

C6-16 Greek column styles: Doric, Ionic, and Corinthian

(colour plate xxvii)

C6-17 Intricate stone tracery designs, Barcelona, Spain

(colour plate xxvii)

the earlier Roman and Greek styles to create impressive churches and public buildings. Brunelleschi designed San Lorenzo, Florence (1419) and the Pazzi Chapel (1430–33), Alberti designed San Andrea, Mantua (1470), and Da Vinci planned a project for a church (1490), as well as creating other similar illustrations in pen and ink (Janson, 1991).

Art Nouveau & Arts and Crafts (1890–1910 AD): William Morris (1834–1896) in England and Charles Rennie Mackintosh (1868–1928) in Scotland reacted against the copying of the past and instead studied nature for inspiration. "Plant-like curves, parabolas and hyperbolas were preferred to straight lines and right angles, squares and rectangles" (Bartuska & Young, 1994, p. 143).

The Bauhaus (1919–1928) was a school of design founded in Germany. It precipitated a revolution in art, design, and architecture that became an attitude of "form follows function." Walter Gropius (1883–1969) was the movement's founder.

Le Corbusier (Charles Edouard Jeanneret, 1886–1965), developed a system in the mid-1920s that he called The Modular. It was based on the golden section, but also on the consideration of the various postures and positions of the human body. Le Corbusier, in his 1927 book, *Toward an Architecture*, wrote the following about geometry, architecture, and human perception:

> Geometry is the language of man. But when determining the respective distances of objects, he invented rhythms, rhythms sensible to the eye, clear in their relationships. And these rhythms are at the root of human activity. They resound in man through an organic inevitability, the same inevitability that inscribes the golden section in children, in the elderly, in savages, in the educated. A module measures and unifies; a regulating line constructs and satisfies. (1927/2008, p. 135)

Non-European Architectural Designs

When studying the architecture of China, India, or Japan, consider the purpose for which the building was designed: for ritual (such as shrines or temples), for nobility (such as castles), or for vernacular purposes serving the ordinary people in their daily lives? Other influences on design include the climate, prospects of natural disasters such as earthquakes (like the powerful quake and resulting tsunami experienced near Japan's coast in March 2011), and the materials that are available for construction. Historically, Chinese architecture from the Tang Dynasty greatly influenced Japanese architecture, for example, elements of Buddhist worship found in Chinese temples were incorporated into the Japanese style. Japan also developed many unique differences and aspects indigenous to Japan as a result of dynamic changes throughout its own long history. The custom of sitting on the floor, as opposed to the Chinese sitting in chairs, changed the use of space. The Japanese use of tatami, which are mats of straw measuring 180 cm × 90 cm to cover the floors, dictated the size and shape of rooms and homes. The variations in seasonal climate and the frequency of earthquakes have also altered common design patterns and forms in that country.

The architecture of the Indian subcontinent was changed over time by the religious influences of Buddhism, Hinduism, Islam, and Christianity, as well as historical forces such as British colonial rule and post-independence social development. Architecture was considered both an art and a science.

Indigenous Architecture

Aboriginal or Indigenous peoples' architecture is in evidence worldwide. The topic of Indigenous architecture covers a broad range of forms and uses, including traditional, ceremonial, or vernacular. It is influenced by the immediate culture and by cultural needs (food and water), region, climate, building materials, and time periods. Indigenous architectural design has, by its very nature, a respectful, symbiotic relationship to the environment.

Many historians have given great credit to the Egyptian, Greek, and Roman cultures for their strong influence on our present understanding of mathematics, art, and architecture. The art and architecture of Indigenous peoples also demonstrate a strong underlying understanding of geometry, astronomy, and mathematics. Details about Indigenous architecture can be found by researching a particular region or culture of interest.

LEARNING EXPERIENCE
Indigenous Architecture Research and Design

- Students can research the Indigenous architecture of a specific region and time period. They should gain a greater understanding of how the shelter met the needs of the people, but also how it worked in harmony with nature. Environmental aesthetics is concerned with the natural and the built environment. Students should consider living spaces and the use of materials that are in harmony with nature. Encourage students to learn from Indigenous wisdom and incorporate this understanding into new architectural designs.

Modern Movements

Modern movements include some of the following:

Modified Historical is a philosophy of design that is influenced by classical Greek and Roman architecture. An example is Charles Moore's (b. 1868), Piazza d'Italia in New Orleans, Louisiana, and I. M. Pei's glass pyramid entrance for The Louvre, in Paris, France (1989).

Modern technology, or "high tech," uses steel and trusses to create uncluttered space for multiple purposes. Jorn Utzon's Sydney Opera House (1957–1973) in Sydney, Australia style is considered **Modern Expressionism**. The opera house is made of tile-clad concrete and pre-cast concrete with spherical formed roofs.

Modern technology has also enabled architects to design environments in what is sometimes called the **International Style**. The "big three" architects who worked in this style were Le Corbusier in France, and Ludwig Mies van der Rohe and Walter Gropius in Germany. The common characteristics of the International Style include an extreme simplification of form, a rejection of ornamentation, and the use of glass, steel, and concrete as preferred materials. The International Style uses simple geometric shapes to create box-like forms wherein roofs and walls are flat and unadorned, reminiscent of abstract sculptures.

Environmentally friendly architecture attempts to be energy efficient and non-toxic, using alternative ways of heating and cooling, such as solar panels. Organizations such as the "Net Zero Energy Home Coalition" in Canada (Hope, 2005) have been founded to research and encourage the design of energy-saving and energy-creating features such as photovoltaic cells in a roof to convert sunlight into electricity.

Douglas Cardinal, a landscape-inspired architect from Alberta, Canada, designed St. Mary's Roman Catholic Church (1967) in Red Deer, Alberta, incorporating certain motifs and meanings originating from his Métis/Blackfoot heritage. He later designed the Grande Prairie Regional College, the Canadian Museum of Civilization, located in Hull, Quebec, the Edmonton Space and Science Centre, and the First Nations University of Canada, located in Regina, Saskatchewan.

Entertainment architecture includes places such as Tivoli Garden in Copenhagen, Denmark, and Brighton Pavilion in London, England. However, in more recent years, building for entertainment is at times associated with flamboyant or elaborate taste, such as the casinos of Las Vegas or the fantastic buildings of Disney World.

Future Architectural Considerations

Future human-designed environments may be very different from what has been designed in the past or to date. There will be a greater need for economically feasible housing in light of the continued increase in world population and the decreasing amount of non-renewable energy resources, such as crude oil and coal. However, with the development of new materials, architects will be free from some of the previous design restrictions.

It will be important for architects to design aesthetically pleasing, yet low cost housing. Housing will be needed for the homeless other than "squatter towns" made of found or discarded materials. Architect Michael Reynolds developed a solution he calls "Earthships," or as others have called them, "tire houses," which can be built by unskilled labour from cast-off tires. Sand and straw are also low cost, easily available materials that have been used to construct buildings.

The potential reduction of "gated communities" due to increased integration of the population and safer public spaces/transportation may also result in new architectural considerations. Another change will be the continued development of environmentally friendly and energy-efficient architecture and living, work, and leisure spaces, biospheres or geodesic domes where families can produce their own food, windmills or

C6-18 Living wall at the University of Waterloo Summit Centre for the Environment, Huntsville, Ontario, Canada (colour plate xxviii)

photovoltaic cells (using sunlight) to produce electricity or passive solar hearing. "Living walls" can now be seen in new buildings, where an entire side of the room from floor to ceiling is covered with plants that draw moisture and nutrients from the sponge-like mats into which they are carefully planted.

LEARNING EXPERIENCE
Architectural Themes and Cross-Disciplinary Research

Experiences should be dependent on students' involvement in other areas of study and/or course requirements. It is important to make valid connections to other subjects whenever possible in order to create meaningful learning experiences for students.

- Investigate architectural themes based on function, geometric shapes, historical periods, cultural influence, and religious influence.
- Investigate and design structures with consideration for elements and principles of composition or design. Consider the proportion of the various parts of the structure and how they relate to one another. Consider areas of dominance and subordination in order to give the structure areas of emphasis. Consider qualities of the structure that repeat, such as shapes, colours, and lines of windows, doors, and pillars.
- Connect the study of architecture as it applies to the study of mathematics, languages, literature (stories and poems), science, social studies, and history.
- Investigate the architecture of a country or culture and how it has changed over time.
- Compare designs from different cultures and periods.
- After studying architectural styles, plan a field trip and have students identify and draw parts of buildings that display architectural features from various periods, styles, and cultures.

- Investigate differences in the way schools are designed and their overall appearance compared to homes, places of worship, libraries, and other public buildings. How did the shape and the way space is used reflect the purpose or function? Were aesthetic considerations involved in designing the building?
- Students can design a three-dimensional model incorporating various architectural features.
- A grid is a set of numbered or lettered squares to enlarge, reduce, reflect, or distort. Find a picture of a building or architectural feature to enlarge. Create a grid on the picture. Use a set ratio and increase the grid lines on a blank piece of paper. Re-create the building larger, smaller, or distorted.
- Reflect on ecological aspects of architecture, considering energy conservation, but also aesthetically pleasing elements and those in harmony with the immediate environment.
- Design a low-cost housing complex. Take into consideration aspects such as the aesthetics (e.g., colour, texture, shape) in harmony with the environment and the importance of personal space in densely populated areas.
- Design houses for various locations, such as on an island, underwater, underground, the moon, or elsewhere in space. Consider different types of structures, such as a geodesic dome, environmental bubble, or Quonset hut.
- Design a house for a little person or for a giant. How do the shape, size, and position of elements of the building change?
- Investigate the natural environment of a specific animal and design a space for it in a zoo. Take a trip to a zoo. Consider all the factors that make specific animals' habitat appropriate for them. What are the limitations? Suggest design alternatives.
- Research the transformation/renovations that took place at the Royal Ontario Museum in 2007, including the addition of the new Michael Lee-Chin crystal entrance. What challenges did the engineers face in implementing

the architect's plans? The new philosophy that fueled the overhaul of the museum involved making museum experiences much more interactive. How was this accomplished? How has the public responded to these major changes in museum layout and philosophy? In general, what controversies often surround the remodelling of heritage buildings?

- Research the Jewish Museum in Berlin, designed by architect Daniel Libeskind in 2001. Why did the architect incorporate slanted walls, slits for windows, and staircases that lead nowhere? What effect might this unusual architecture have on visitors, and why might this effect be intended by the architect?

C6-19 Jewish Museum by Libeskind in Berlin, Germany (colour plate xxviii)

LEARNING EXPERIENCE
Drawing Buildings and Landscapes

- Study homes in the neighbourhood or around the school, such as bungalows, bi-levels, duplexes, fourplexes, apartment buildings, high rises, and so forth.
- Using footsteps as a ratio (for example 10 steps = 1 mm) create a grid and draw a map of the neighborhood, city street, or block. Take a walk around the neighbourhood. Draw rough sketches of several buildings.
- Have students use sight measurements to draw their own home in proportion. Find the size of a part of the house, such as the height of the front door, by measuring the door using the end of a pencil or ruler (it may help to shut one eye). Use this measurement as the ratio in order to measure other parts of the house, such as its height and width.

- After drawing the house, create a ratio using these measurements and build a three-dimensional model of the house. Draw the plans to scale using a ruler. Build the model using various construction materials, such as heavy paper, cardboard, boxes, hard plastic, found objects, recyclable material, clay, papier mâché, or any combination of materials.

C6-20 Parthenon three-dimensional model structure (colour plate xxviii)

- Using a grid, draw a picture of a building by viewing it upside down. This becomes an interesting exercise in drawing—what students actually see rather than what they believe they see (or have memorized as symbol).
- Draw a grid with non-permanent felt markers on a window or piece of Plexiglas. Draw what is seen on the other side of the window. Re-draw on paper by measuring the various parts. Re-draw an enlargement or reduction using a grid.
- Draw a building in a landscape and overlap the building with the landscape. When deciding on the placement of the horizon line and details in the foreground, middle ground, and background, consider the rule of thirds from photography (i.e., dividing the canvas/page into thirds both length-wise and width-wise, and treating the four intersection points as potentially strong focal points for objects).
- Draw a building quickly in 5 to 10 minutes, estimating the proportion of the parts. Draw the same building taking 45 to 90 minutes, sight measuring size and proportion, and adding detail.
- Draw the building using wax crayon on a piece of Plexiglas. Place a piece of drawing paper over the Plexiglas crayon drawing. Using a medium hot iron, transfer the image onto the paper.
- Draw a building on Plexiglas and use tempera or watercolour paint to

complete the image. Place a piece of wet drawing paper over the Plexiglas painting. Use hands to rub over the paper to transfer the image. This is called a monoprint.

- Draw detailed sections of a building; for example, draw pillars, arches, doorways, receding windows, moulded window frames, elaborate roof edgings, overhangs, covered porches, and balconies.
- Choose a building or picture of a building. Make note of the geometric shapes and the types of angles. Are the angles complementary (add to 90 degrees) or supplementary (add to 180 degrees)? Draw the building using the shapes and angles.
- Capturing city lights: Visit a spot where you can view a city, harbour, or street. Depict this view using dark tones of paint, conté, or chalk, and light tones for the bright spots of natural light (e.g., moon, stars) or artificial light (e.g., street lights, building interiors, vehicles). Look at the city view shown in Vincent Van Gogh's (1853–1890) painting *Café Terrace at Night* (1889). His other paintings of starry skies and cast light include *Starry Night over the Rhone River* (1888) and *Starry Night* (1889) depicting a night sky with stars and moon, hills, and a small town.

C6-21 Drawing buildings from enlarged photos

LEARNING EXPERIENCE
Drawing a Building Using Coordinates

> *Mathematical shapes can always be reduced to numbers—which is how computers handle graphics. Each tiny dot in the picture is stored and manipulated as a pair of numbers: how far the dot is along the screen from right to left, and how far up it is from the bottom. These two numbers are called coordinates of the dot. A general shape is a collection of dots, and can be represented as a list of pairs of numbers.* (Stewart, 1995, p. 6)

Coordinate geometry involves two sets of intersecting lines drawn perpendicular to each other. The position of any point can be given by its distance from each of the axes lines (i.e., horizontal *x*-axis and the vertical *y*-axis). Numbers give a position on the horizontal and vertical axes on a graph. The coordinate plane divides a space into four quadrants with a vertical and a horizontal line that intersect. This system is known as the Cartesian plane in mathematics and is named after the French mathematician and philosopher, René Descartes, who developed the idea while convalescing in a bed as a sick child (watching a fly on the ceiling, as the story goes, and wondering how he might describe the position and movement of the insect using a standard grid). Coordinate systems can be used to draw many things, including buildings.

▣ Have students use the following procedure to draw a building using the coordinate system and sight measurements:

1. Sight measure the distance from the top of the building (e.g., peaked roof) to the ground level.
2. On grid paper, mark the top and bottom of the building on the vertical. Make a vertical line on the right-hand side of the paper, joining the marks you have made at top and bottom.

3. Sight measure the distance from one end of the building to the other on the ground level. Place the points on the grid.
4. Sight measure the distance from the top to the bottom of the building on one side and then the other. Place the coordinates on the grid.
5. Sight measure the distance from one end of the building to the other under the roof. Place the coordinates on the grid.
6. Sight measure the various architectural elements as they are in relation to one another and place them on the grid.
7. Connect the points (coordinates) to create a drawing of the building.

Creating the Illusion of Depth

Artists have developed many techniques that give the illusion of depth on a two-dimensional surface. Some of these methods include overlapping, relative size, position of the object on the picture plane (a picture that attempts to give the illusion of depth should have a foreground, middle ground, and background), level of detail, and tonal contrast. Becoming aware of these techniques can help the artist create the appearance of depth in a piece of art or to avoid the illusion of depth when a flat surface is desired.

Overlapping: This is a basic technique used to create the illusion of depth. One shape is partly hidden from view by another shape; for instance, a tree placed behind another appears farther away.

Size: Varying the size of objects or shapes is another technique. Smaller objects appear farther away; for instance, in a forest, larger trees appear closer and smaller trees appear farther away.

Level of detail: Greater detail can be seen on objects that are closer; for instance, the veins on the leaves of trees are visible up close. However, from farther away, the veins cannot be seen, or appear blurred.

Position on the picture plane: Objects higher on the page appear farther away because they are positioned closer to the horizon line. In a landscape painting where trees are painted closer to the horizon line, they appear to be in the distance. Trees lower on the page appear closer.

Colour: The colour of objects that are closer are more vivid. Colours are less distinct, or greyer, on objects that are farther away. In landscapes, objects in the distance often appear blueish, taking on the colour of the sky ("atmospheric perspective").

Linear Perspective

Linear perspective is one of the most common techniques used by artists to give the illusion of depth. The rules of drawing objects in perspective, developed as a high science during the Renaissance era, are a mixture of geometry, arithmetic (numbers), and visual perception in determining a system of lines. Converging lines are used as guides in drawing objects that appear to recede. For example, one-point perspective might show a narrowing road meeting at a point on the horizon line, where the sky appears to meet the earth.

Children develop at different rates and are usually not ready to work with linear perspective until they are about ten or twelve years old. If you feel your students would be able to understand the methods of creating linear perspective, introduce them to terms such as horizon line, converging lines, vanishing point, one-point perspective, and two-point perspective. Perspective should be taught step by step and with great patience. Once the concept is understood, children become very inventive in using the technique.

The horizon line is always at eye level for the viewer or artist. When lying

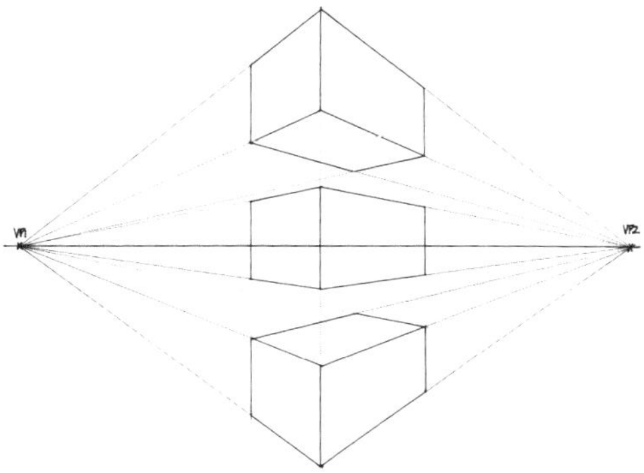

C6-22 Drawing of boxes with two-point perspective

C6-23 Drawing with one-point perspective

down on the grass looking up, a greater amount of sky is seen. When flying in a hot air balloon looking down, more land will be seen. Converging lines are the lines that meet at a vanishing point, and the vanishing point is a point on the horizon line. One-point perspective lines converge at one point, such as the lines of a railway track that meet on the horizon line. Two-point perspective has lines that converge toward two points on a two-dimensional, flat surface.

LEARNING EXPERIENCE
One-Point Perspective Drawing

- Collect a number of boxes of different shapes and sizes, at least one for each student. Have students hold and observe a box in various positions (above and below the horizon line, at eye level, and to the right and left). Illustrate one-point perspective on the horizon line toward the left and right (at eye level), and above and below the horizon line. Discuss and demonstrate each idea (i.e., model these drawings on the chalkboard, overhead projector, or interactive whiteboard) one step at a time. The height of the box can be measured and the depth of the box can be visually estimated and drawn on the linear perspective line.

LEARNING EXPERIENCE
Two-Point Perspective Drawing

- When you are demonstrating the use of two-point perspective, start by focusing on the vertical edge of the box that is positioned closest to the viewer (rather than on the flat surface or side of the box for one-point perspective).

1. Look at the box at eye level. Place a perpendicular line across the horizon line to represent this box edge that is closest to you.
2. After placing two vanishing points on the horizon line, one at each end of the paper, draw the converging lines from the endpoints of the perpendicular line (box edge closest to the viewer) to each of the two vanishing points.
3. Estimate the length and width of the box, draw two new lines parallel to the first perpendicular line (front box edge) to represent the back two edges of the box.
4. Now complete the two-point perspective image by drawing lines from these new vertices (corners) to the opposite vanishing point (i.e., the one farthest away from each vertex) in your picture.
5. Draw the underside of the box if it is above the horizon line. Draw the top side of the box if it is below the horizon line.

- Use this technique to demonstrate various points of view. Encourage students to look at the box from many points and views, such as an ant's eye view, bird's eye view, front, back, top, sides, and angles. Draw the box from each point of view. Students can exchange various sizes of boxes with each other to measure and estimate.
- The boxes can be depicted in invented or imaginary environments using pen and ink, decorative lines and shapes, or a particular colour scheme.

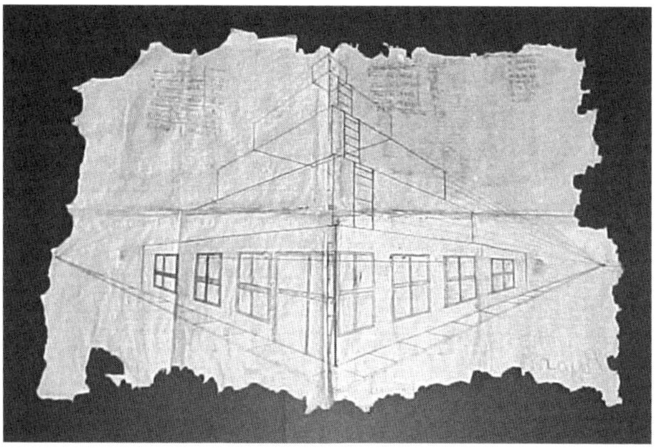

C6-24 Drawing of a building with two-point perspective

LEARNING EXPERIENCE
Exploring Illusion of Depth Techniques

- Deconstruct reproductions of paintings and drawings in which artists have used illusion of depth techniques to depict landscapes, buildings, and bridges. Consider where the artist or photographer was situated. Where would the artist's eye level be? Where was the vantage point: high on a hill, in an adjacent high building, on the ground, or in a valley? Understanding this will help determine where the horizon line on the picture would be. Consider where the one or two points would be located, on or off the picture plane. It may be helpful to use a pencil held horizontally in order to determine the lines of perspective. Draw in the lines of perspective.
- Encourage students to draw or paint pictures incorporating a variety of techniques to create the illusion of depth on a two-dimensional surface.
- Take a camera and a viewfinder along on a walk. Use the viewfinder when selecting the area or composition to photograph. With the aid of a view-

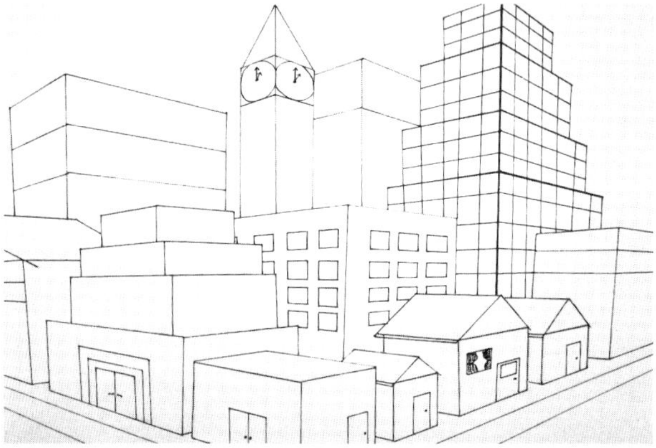

C6-25 Drawing of city block with two-point perspective

finder, choose a spot two to three kilometres in the distance. Draw the landscape as viewed from a distance. Consider the ways of depicting depth: the visual aspects that give the scene the sensation of depth, such as size and overlapping, greying of colour, detail, and proximity to the horizon line.

- From one point of view, draw a row of houses down a street. The buildings farther away appear small. Notice the visual features of depth. Locate where the horizon line and vanishing point should be.
- Overlap a drawing of buildings with a drawing of a landscape.

Homes in Students' Own Environment

View several homes and other buildings from pictures or real field trips around the neighbourhood block, around a subdivision, county, town, or city. Study blueprints or floor plans from commercial builders from show homes and open houses. Consider the location of the building alongside the plans.

Upon close observation and discussion, decide whether the architects considered the human need for aesthetics (form), or did they mainly consider cost efficiency and/or required elements of the building (function).

Questions of Form: Does the design fit with the neighbourhood or local culture? How do light, colour, and texture affect the space? What is the visual impact of the building? What kind of emotional response, if any, might a person have while using the designed environment? Does the design work in harmony with nature?

Questions of Function: Is the design suitable for its environment (e.g., withstanding hurricanes, earthquakes, hot/cold climates)? Is the roof shape determined by the climate or other natural forces? Does the design maintain privacy in dense housing areas? Does the design take into consideration the size of an average human and how people will move comfortably and safely through the building or related spaces?

Look at maps of new and old suburbs. Does the layout of buildings consider a regular occurrence of open space with vegetation, such as grass, shrubs, and trees? Does it work in harmony with nature? It might be possible to invite architects or building contractors to visit the class to discuss their work with regard to mathematics, aesthetics, and the natural environment. Interested students might be partnered with a mentor or take part in a job shadow arrangement to gain first-hand knowledge of such careers.

LEARNING EXPERIENCE

Form versus Function Analysis of Structures

- Have students investigate the shapes, sizes, use of space, building materials, and structural support (e.g., shapes of beams, columns, and roofs) that are used in a particular architectural design. Which aspects of the building relate primarily to form and which primarily to function?

LEARNING EXPERIENCE
Build Your Dream Home

▣ After viewing and discussing a variety of buildings as designed in their environments, students can begin planning their own dream home. This project can be done individually or in small groups. Forming small groups encourages students to work cooperatively and inventively, using their combined talents and abilities. What type of house will it be? Will it be a bungalow, a two-story, a split-level, duplex, townhouse, apartment, or another structure? Consider the following factors:

- Location, size, shape, and land contour or terrain, soil, or other foundation features of the building lot. What size and shape of home would be best suited for the lot, the neighbourhood homes and other designed structures, the natural environment, and climate?
- The estimated cost of the house
- The estimated cost of the house per square metre
- The total size of the house in square metres (Usually only the main and upper floors are added to get this total, although sometimes the basement is also included in total floor space.)
- The number and size of rooms to be accommodated in the house. These could include a living room, family room, den or study, media/entertainment room, master bedroom, nursery, and other bedrooms, bathroom/s, dining room, kitchen, laundry room, exercise/art/craft/sewing/work room, storage area, attached or detached (one or more) car garage. The rooms and the function of specific rooms are a personal choice.
- The dimensions of the walls, including the height from floor to ceiling

- The number of floors the house would have (e.g., one, two or more, or split level)
- The best location for the stairways and hallways
- The location and size of doors, windows, turrets, porches, and balconies
- The shape of the roof and additions or details on the roof (e.g., chimney or gables)
- Outdoor spaces such as vegetable, flower, or rock gardens, pools or other water features, or climbing gyms, trampolines, slides, or other play structures
- The front entrance of the home, including sizes and shapes of windows, doors, overhang, and roof
- The exterior finish of the house (e.g., brick, wood, siding, plaster)
- The roof finish (e.g., shingles, metal, clay, thatch)
- The price of the various materials needed in order to make a scale model or to determine the cost of constructing the designed home

- Begin by drawing rough, two-dimensional layouts of a building. Fine-tune the drawing to scale with a bird's eye view for each floor. Find the square metre size of each room. Determine the size of the total floor plan in square metres (area = length × width). Decide on the size of the house and the size of the various rooms. Decide on the scale for the floor plan (for example, 1 cm = 1 m).
- A three-dimensional scale model can be planned and executed as well. Various media and materials can be used to construct the house model, the creation of furnishings, and landscaping. In order to view the inside of the home, the roof should be made to be removable. All the jobs require students to measure and build to scale. Mathematical concepts include ratio and proportion, measurement, area, perimeter, estimation, data collection, two- and three-dimensional drawing, and analysis.
- If students are working in groups, they may choose various jobs. These could include the architect who designs the blueprints, the construction

contractor who oversees the construction, the interior designer who oversees the decorating and furnishings, and the landscaper who oversees the landscaping phase for the project.

- After the house is designed and built, the next step (before or after interior furnishing) is consideration for the landscape. Decide on the area for the landscape and the perimeter for the fencing and sidewalks. The landscape design will include paved areas such as a driveway, walkways and a patio, grass, trees, shrubs, a garden, and fencing. The design should have consideration for lines, shapes, textures, and colours. What is the budget for the landscaping? Price the various aspects of the landscape design from garden and hardware store catalogues.
- The interior designer is responsible for furnishing and decorating the interior. Selections include floor and wall coverings, ceiling style and colour, appliances, shelving, tables, chairs, beds, etc. Furniture and hardware store catalogues can help in the selection of furnishings and furnishings and the cost.

Drawing Interiors

Interiors can be drawn using one- or two-point perspective. Hallways work well when drawing one-point perspective. A corner of a room works best with two-point perspective.

LEARNING EXPERIENCE
Rendering Interiors with Perspective

- As a first step, use sight measurements to draw a bird's eye floor plan of a room to scale.
- Draw a corner of a room using one-point perspective.

- Draw a corner of a room using two-point perspective.
- Shadows and lighting: How does lighting, both natural and artificial, affect the feeling in the room? How does special-effect lighting change the ambiance of a space?

Streets, Playgrounds, and City Planning

Every town or city has a planning department that investigates and designs various transportation systems, such as streets, highways, overpasses, and light rail transit. They also plan where government offices, libraries, and other public buildings will be constructed. They assist in, but also set restrictions for, the development of plans for residential, recreational, and commercial sites.

Working with landscape architects, landscaping companies, and community members, the city planning department attempts to design and develop landscapes and public spaces that best serve the needs and preferences of all stakeholders within a local community. Landscape architects often use plants that are native to the environment, or at least able to withstand the local climate. They also give consideration to the elements of art/design in their landscape creations.

C6-26 Streets and city planning (colour plate xxviii)

LEARNING EXPERIENCES

Invented Landscapes and Fantasy Architecture

- Select an area that facilitates several uses, tour the area, and develop a map. What are the strengths, weaknesses, and interesting parts of this area? Are there historical sites or recreational areas that need some enhancement in your community? What could be altered or added to make the area a more pleasant environment?
- Select a picture of a building that was designed by a well-known architect such as Frank Lloyd Wright, Le Corbusier, Jorn Utzon, Douglas Cardinal, Arthur Charles Erickson, or a local architect of lesser renown. Draw the building in detail by measuring the parts or using a grid. Invent a landscape in which the building would be most suited.
- Create architectural models based on a floating city, a walking (no vehicles) city, a city under the sea, or an orbiting satellite (i.e., space-based) community.
- Design an island or an imaginary small continent. Draw a contour (topographical) map illustrating mountains, hills, valleys, rivers, and lakes. Decide on the location of this landform—its longitude and latitude. The seasonal weather will have a direct influence on the flora, fauna, and types of materials available and will also determine the shelter needs for the hypothetical inhabitants.

C6-27 Students working on clay building façade designs

Bridges

Bridges are not normally regarded as art or architecture, but rather as engineered structures. However, they do become an integral part of the environment and, as such, require aesthetic considerations like any other designed building or public space. Ancient bridges were usually a combination of fallen trees and stones placed over gaps, valleys, streams, or rivers. A bridge can be a pathway, a railway, a motorway, or a barricade such as a dam, but is rarely an actual building that accommodates people within. The Romans built bridges with arches so as not to impede the flowing water.

"The use of steel increased the possibilities of the suspension bridge, which hung on enormous cables from pier towers as near as possible to the banks, [and which] provided the longest and most impressive spans over water" (Martienssen, 1976, p. 93). Engineers extended the potential of bridges with the use of iron, steel, and concrete using advanced mathematical calculations. The architects and engineers need to consider many things when designing these structures. A primary concern in designing bridges and motorways is stability. Will the structure be stable or unbalanced considering the terrain, soil conditions, and potential natural disasters? Does the structure work in harmony with the natural environment or the surrounding constructed environment?

LEARNING EXPERIENCE
Exploring Bridges

- Examine photographs of many types of bridges with arches, cantilevers, columns, beams or suspension supports. How do these make the structure stable? Sketch, name, and label these bridges. Make note of the location, immediate environment, climatic conditions, and recurring forces of nature.

- Examine each bridge with the following considerations: What parts of the bridge are under compression (i.e., the parts that press together to increase support)? What parts are under tension (i.e., the supports that pull apart to increase support)? How has the bridge been made stable, firmly positioned, or supported against shear force that may be caused by natural elements? What other ways is the bridge supported?

C6-28 Centre Street Bridge over the Bow River in Calgary, Alberta, Canada (colour plate xxix)

- Does the bridge work in harmony with the natural and/or constructed environment? Structure your bridge critique in terms of positive/negative/interesting elements.
- Are there bridges in the neighbourhood? Draw pictures of these bridges, especially the shape and types of support. What is the main use of the bridge: pedestrian, train, bike, vehicle, or dam? Describe and draw the methods and materials of construction.
- Bridges can produce an emotional response from the viewer or user. Imagine sailing under the Golden Gate Bridge (1933) in San Francisco, California. Consider driving over the suspension Brooklyn Bridge (1869-

C6-29 Lion's Gate on Chain Bridge over the Danube River in Budapest, Hungary (colour plate xxix)

C6-30 Tower Bridge over Thames River in London, UK (colour plate xxix)

1883) in New York, the Lions Gate Bridge (officially known as First Narrows Bridge, 1937–38) in Stanley Park, Vancouver, British Columbia, the Railway Trestle Bridge (1909) in Lethbridge, Alberta, or the Oresundbroen Bridge (2000) between Denmark and Sweden. The Tower Bridge (1886-1894) in London, England, has been the inspiration for many stories and myths over the years, and it is impressive to walk over the Sydney Harbor Bridge (1932), known as the "Coathanger," in Australia. Are they primarily functional in their design? What type of bridges are these and what construction materials were used? Did they require conscious decisions on the part of the architects and engineers to create an emotional impact, or to be aesthetically pleasing and at peace with the environment? Look at pictures of bridges and design a dream bridge.

Summary

In this sixth chapter, we have explored the connections between mathematics, visual arts, and the designed environment. Examples of European, non-European, and Modern architecture styles have been described. Learning experiences in this chapter have students observing their built environments, mastering different forms of perspective drawing, analyzing and discussing the tension between form and function inherent to architectural projects, researching bridges, designing landscapes, and creating their own dream home.

7 Mathematics and Art in Designed Objects

Since man lives in a series of environments that are partially or totally man-made, it is fair to ask: Where does environmental form come from? The origin of man-generated physical form . . . lies in the interaction of human needs, attitudes, and values with natural laws and materials. Both man-made and natural objects can be read as a record of the forces that introduced them. (Preble, 1973, p. 356)

Introduction

The visual arts have played a significant role in societies, wherein they have enriched rituals, ceremonies, and other special events. They have also played a significant part in educating and socializing humans in terms of cultural expectations, social history, and contemporary thought. This chapter discusses the connection between mathematics and designed objects. These objects include architectural details and decorations, two- and three-dimensional shapes, cultural circles, calendars, time and money representations, and a number of games/activities. Also included is a discussion regarding how mathematical concepts are often an integral part of the work of some contemporary visual artists. Learning experiences for mathematics and the visual arts, as well as key concepts and images for process and product, appear throughout the chapter.

Territoriality serves privacy needs in two major ways. First, it allows us to be who we are—to raise our families, take care of biological needs, withdraw from community interaction, develop ourselves. And, second, when we modify our homes, personalize them, mark them with symbols of our selves, we are using territoriality to assert our individual identity as well as our allegiance to a neighborhood, an ethnic group, a taste culture, or a class. (Kron, 1983, p. 29)

Mathematics and visual arts play a significant role in societies. Since prehistoric times, be it in caves, tents, or more substantial buildings, humans have been concerned with the embellishment of their dwellings. Decorating and reorganizing a space can be referred to as "marking" or "personalizing." It is also a way for the cultural group to educate and socialize members, especially the young, in expectations, rituals, ceremonies, spirituality and religion, history, and current thought. How humans divide spaces and decorate the interiors can often reflect social class, status (or sought-after status), values, gender, ethnicity, and age.

C7-1 Colour façade of the W. C. Cameron building in Victoria, British Columbia, Canada (colour plate xxix)

Designed interiors are spaces within the walls of a constructed environment. The size and shape of the interior, as well as the objects and their arrangement reflect the function of the space. Hall (1966) developed the concept of "proxemics." He determined that humans exist in basically four zones, or spaces, which include intimate, personal, social, and public. Added

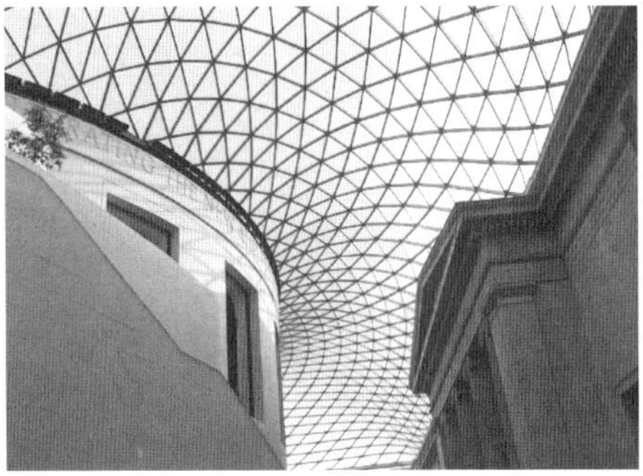

C7-3 British Museum interior space in London, UK

C7-2 Christ Church Cathedral interior space in Fredericton, New Brunswick, Canada (colour plate xxix)

to this list might be religious, spiritual, or meditative spaces. Consideration of these spaces is required when designing for size, function, and objects.

Objects such as chairs and tables are designed for specific functional and behavioural purposes, and are influenced by available tools and materials. The measuring of human body parts in static positions and the understanding of the function of the parts and how they move in three-dimensional space is called anthropometrics (Bartuska & Young, 1994). Ergonomics is the science of making the work environment safer and more comfortable for workers using design and anthropometric data.

LEARNING EXPERIENCE
Measuring and Observing a Room Interior

- Measure the classroom. Determine the area of the room in square metres. Divided equally, how much space does this give each student and the teacher working in the classroom at one time?

- Identify the shapes that can be found in the classroom. What elements of composition (line, colour, shape, texture, tone) or mathematical terms (two-dimensional shapes such as squares, rectangles, circles; three-dimensional figures such as rectangular prisms, rectangular pyramids, spheres, cylinders, cones) can be used to describe objects in the classroom?

LEARNING EXPERIENCE
Form and Function in Design

- Discuss furniture and other useful objects found in the home and ways in which the function of each object has influenced its shape by asking the following questions: What is meant by the word functional? Why should a chair, or table, or bookshelf be functional?
- The design adage "form follows function" means that the chair should not only be aesthetically pleasing, but also have stability. Observe and draw several types of chairs. Consider what features assist in creating stability. Do these features add or take away from the aesthetics of the chair?
- Redesign other functional objects. How have you made them more aesthetically pleasing?

C7-4 Redesigned egg and holder in art period style

LEARNING EXPERIENCE

Drawing Interiors

- The meeting place of form and function is nowhere more significant and conspicuous than in the designed spaces in which humans live, work, and play. That there is something intriguing about designing and defining a three-dimensional space first becomes evident in the focused play of a child. Playing with miniature buildings and public spaces/communities using Lego, Ello, three-dimensional immersive video games, and so forth is popular among children and continues to be highly engaging for adults as well. In this activity, students carefully observe and document a floor plan for a selected room and apply linear perspective. Using sight measurement, they then draw a floor plan of a room to scale.
- Draw a corner of a room using one-point perspective.

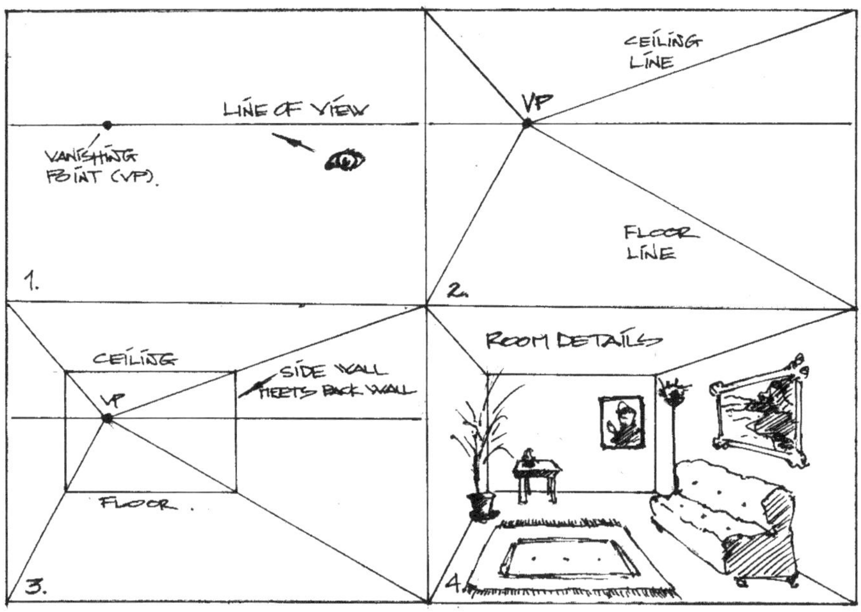

C7-5 Room interior rendered with one-point perspective

1. To begin, locate the centre of the room according to your eye level. Position the vanishing point on a horizontal line that approximates your line of view.
2 Draw the floor and ceiling lines converging to this vanishing point.
3. Draw two vertical lines to represent where the sidewalls meet, and then horizontal lines to represent where the far wall meets the ceiling and the floor.
4. Finish by adding shaded room details such as furniture, shelves, and pictures.

C7-6 One-point perspective as seen in a street scene

▣ Draw a corner of a room using two-point perspective.

1. Begin by drawing the horizon line to approximate your line of view. Add two vanishing points on the horizontal line.
2. Draw a vertical line representing the meeting place of two walls.
3. Draw construction lines (i.e., lightly drawn lines used to guide the construction of other key lines/objects in perspective drawing) that converge to the two vanishing points.
4. Use two more vertical lines to denote the approximate meeting places of the walls.
5. Add objects by continuing to use the two-point perspective convergence of lines for parallel edges.

LEARNING EXPERIENCE

Diorama Creation

- Create a diorama of an interior. A diorama has a background and objects and images in the foreground and middle ground placed so that all are simultaneously in view.

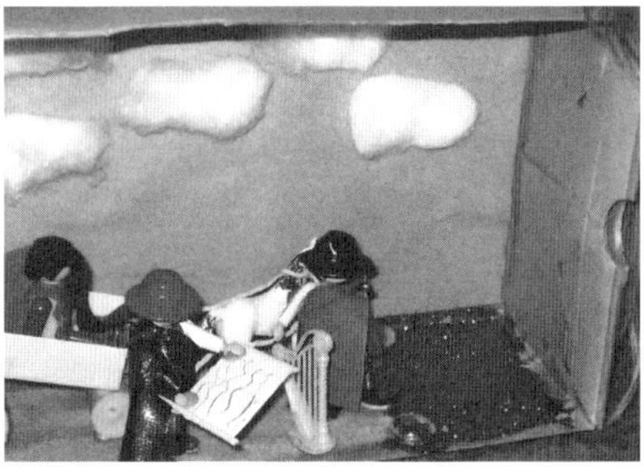

C7-7 Shoe box diorama interior

C7-8 Christmas nativity diorama

LEARNING EXPERIENCE

Still Life

- Using a viewfinder (see page 71), draw an enlarged still life of found object(s) (e.g., chair for giant) portrayed in a fantasy-type space. In order for this to be successful, several ways to create the illusion of depth should be incorporated into the drawing (e.g., converging lines, colour theory, overlapping of shapes, position on page, amount of detail).

Nature as Inspiration

Designers of articles for everyday life—both functional and non-functional articles—often gain their ideas and inspiration from nature. Shapes, colours, and patterns are found everywhere in nature. Artists, architects, and interior designers have looked at natural forms for inspiration and have often modelled their human creations based on aesthetics found in nature. One example is the number approximation for phi (1.61803), and the related ratio of "phi to 1," which are found extensively throughout the natural world in plants, animals, and even human beings (e.g., the five-pointed star in an apple cross-section, a nautilus shell). See chapter 3 for more details on the golden ratio.

LEARNING EXPERIENCE
Exploring Patterns in the Environment

- Look at patterns in the natural and the constructed environment. Students can reproduce, create, repeat, and combine regular, irregular, symmetrical, and asymmetrical patterns and shapes by creating stamp prints or by using cut-out coloured paper.
- Research various architectural features and consider what natural object or phenomenon might have inspired the artist or architect.

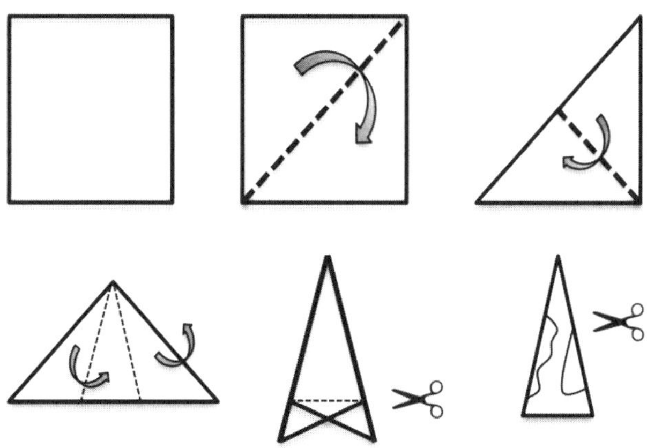

C7-9 Snowflake construction

- Observe the mirrored images of butterflies, moths, and other insects to study symmetrical patterns.
- Patterns can be formed by folding and cutting paper to create symmetrical images such as butterflies or snowflakes.

C7-10 Paper snowflake designs

- Paper snowflakes are designed by (i) folding a square piece of paper in half diagonally; (ii) folding this in half diagonally again; (iii) folding the remaining triangle shape in thirds, one side in front, the other in back; (iv) trimming the extra piece off of the bottom; and (v) cutting out design shapes from around the edge of the triangle. These shapes are then duplicated in the design when the paper is unfolded.
- Make several snowflakes of different sizes, cut with different angles and curves. For display purposes, you might consider the following: Hang groupings of snowflakes about the room (glued onto a sheet of wax paper), create a mobile with the snowflakes, tape them onto the windows, or mount them on darker coloured paper and display. Discuss elements and principles of design that are present in the various snowflakes, especially pattern, positive and negative shapes, balance, repetition, symmetry, and asymmetry.

Golden Rectangles, Triangles, and Spirals

Golden spirals are formed by drawing quarter circles within the iterated, or "whirling" squares of the golden rectangle. The compass point is always placed on the inner (towards the middle of the rectangle) corner of each square and then the quarter arc (one fourth of the circle, or one quadrant's worth) is then sub-

C7-11 Golden rectangle shelves

tended/drawn to join the next part of the spiral. When all connected, these arcs (parts of circles) drawn within the golden rectangle squares form a golden spiral which has often been used as a design template in works of art. Alex Colville, for example used a double golden spiral with the two focal points (the visual vortex of where the spiral leads) as key design spots for objects (the centre of the head of a person sitting on the dock, and the centre of the head of a person entering a docked boat) in his painting *Embarkation* (Jarvis, 2002, p. 19).

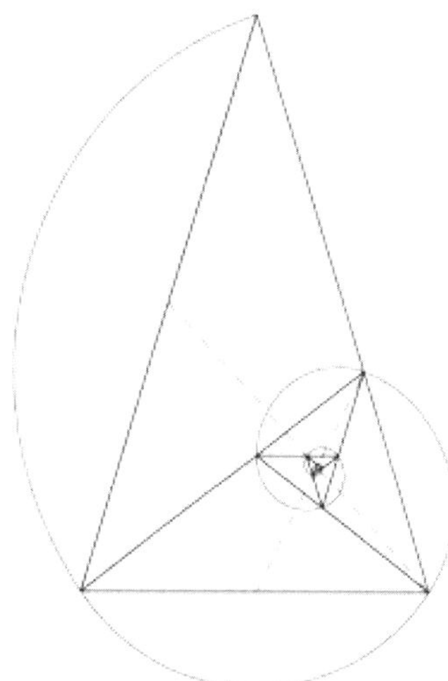

C7-12 Golden triangle and the logarithmic spiral

Golden triangles, or "sublime triangles," are isosceles (two equal angles and two equal sides) and have the special property of having side lengths that are in golden proportion to one another (i.e., when the longer side is divided by the shorter side, the quotient approximates phi, or 1.618033 . . .). This results in the two similar angles being 72° and the smaller angle being 36°. To construct a logarithmic (or equiangular) spiral from the golden triangle, one simply bisects (divides in exact halves) the two equal base angles, then uses the new intersection point of these bisectors as the corner of a new (internal) golden triangle. Like with the golden rectangle/spiral, this bisection process can be repeated infinitely (as small as one can draw), creating an infinite (neverending) number of golden triangles. The logarithmic spiral can then be drawn by connecting arcs (sections of circles) that are drawn with a compass that is repeatedly centred on the inner vertex of each subsequent golden triangle.

LEARNING EXPERIENCE
Creating a Golden Rectangle and Triangle

- Create a golden rectangle (see Jarvis, 2008; also described in earlier chapters of this text) and then accurately construct the internal repeating squares and the associated circle segments to form a golden spiral.
- Create a golden triangle by measuring the three key angles with a protractor (36°, 72°, 72°), and then construct the internal repeating golden triangles and associated circle segments to form a logarithmic spiral.
- Artists and architects have used the golden ratio and rectangle as design frameworks for creating paintings, objects, buildings, and public spaces. Use one or both of the above creations (golden rectangle or triangle) as a design template for a work of art, with key focal points occurring at the heart of the golden and/or logarithmic spirals.

The Bauhaus School of Design

The Bauhaus, founded in 1919 in Germany, was a school that combined crafts and the fine arts, and was famous for the approach to design that it publicized and taught. It precipitated a revolution in art education and design, and the Modernist influence of the program is still felt today in attitudes about art, design, architecture, and crafts (Whitford, 1984). More than a style of architecture and artifacts, it became an attitude of design.

The school existed in three German cities (Weimar, 1919–1925, Dessau, 1925–1932, and Berlin, 1932–1933), under three different architect-directors: Walter Gropius (founder, 1883–1969), Hannes Meyer (1899–1954), and Ludwig Mies van der Rohe (1886–1969). Other members included Albers, Bayer, Breuer, Itten, and Moholy-Nagy. The design students at Bauhaus were trained in the language of form. "Determined to reject all forms of applied ornament and pretentiousness, they would allow only one determining factor in their search: function" (Rowland, 1990, p. 13). Preliminary courses included the study of the Old Masters, optical illusions, and the work of artists such as Kandinsky. Moholy-Nagy incorporated his interest in the "Mazdaznan Cult," in which students were encouraged to meditate, go through group breathing and rhythmic exercises, and depict emotions abstractly (Rowland, 1990). Albers directed his students to work with paper and scissors, folding and cutting, learning economy of size and shape, transforming flat sheets of paper into three-dimensional structures (Rowland, 1990).

The school was closed by its own leadership under pressure from the Nazi regime in 1933. The Bauhaus greatly influenced the form and function of decorative objects, tableware, lights and lamps, furniture (tables, desks, chairs), toys, textiles, graphics, painting and sculpture, stage design, and architecture.

LEARNING EXPERIENCE
Tiled Floor and Bauhaus Objects

- Using one-point or two-point perspective, have students draw a tiled floor with at least two Bauhaus-style objects on it. Students might research the Bauhaus style and either copy existing pieces of furniture or create their own designs for this drawing.

Crests, Trademarks, Logos, Medals, and Flags

Crests, logos, and trademarks include images and words that brand or label such things as packaging, products, businesses, sports, and families. Every culture has had artists design these for many reasons and occasions. Crests were named for their traditional placement on the top of a helmet as on a blue jay's head. Today crests are often attached to cloth or topcoats and jackets. Medals and medallions are generally sculpted or moulded to depict an image and are designed and given as recognition of an accomplishment. Flags are generally made on rectangular fabric symbolizing a country or event with colour, line, and shape. Images, colours, shapes, objects, figures, animals, fantasy animals, birds, insects, instruments, and tools are carefully chosen with consideration and incorporated into the designs.

LEARNING EXPERIENCE
Designing Logos, Trademarks, Medals, and Flags

- Students can investigate an event such as the Olympic Winter Games. Every four years, the hosting country develops a logo for its Olympic games. Designers also develop trademarks and medals. Choose or develop a logo, trademark, medal, or flag for an event or imagined event. Develop sketches of images to be used. What elements and principles of composition need to be considered? What mathematical concepts are involved in the design decisions and production?

LEARNING EXPERIENCE
Name and Numbers in Perspective

- After discussing the terms and viewing two-dimensional images that illustrate linear perspective, have students draw their name, selected letters, or numbers. The illusion of depth can be created by converging all the edges to one vanishing point. They can colour, decorate, or paint their names in an invented environment, or incorporate symbols into the design to represent personal interests or characteristics. Students may also incorporate other methods of depicting the illusion of depth such as detail, size, overlapping of shapes, vivid colour, and position of objects on the page.

C7-13 Student name in one-point perspective

Creating with Two-Dimensional Shapes and Three-Dimensional Solids

The following learning experience combines geometry, compass skills, construction, composition, problem solving, and art materials to create original artworks. The construction of these forms requires students to measure angles, lengths, and surface area. Shapes (two-dimensional) and solids (three-dimensional) that can be constructed include

Polygons: triangle, quadrilateral, pentagon, hexagon, heptagon, octagon, decagon
Prisms and pyramids: rectangular prisms or pyramids, triangular prisms or pyramids, circular prisms (cylinders) or pyramids (cones)
Platonic solids: tetrahedron, hexahedron, octahedron, icosahedron, dodecahedron

LEARNING EXPERIENCE
Creating 2-D Nets for 3-D Solids

- First, students must measure the relevant angles and sides, enlarging and drawing the two-dimensional pattern, or net, for the desired three-dimensional solid. The next step includes drawing and painting the surfaces in such a way that the colour and other design elements enhance a visual sense of movement. Colour schemes can be chosen for the visual effect. The final stage is the cutting, gluing and assembling of the pattern into a three-dimensional form. Many of the three-dimensional forms, or solids, can be built to create stabile or mobile structures.

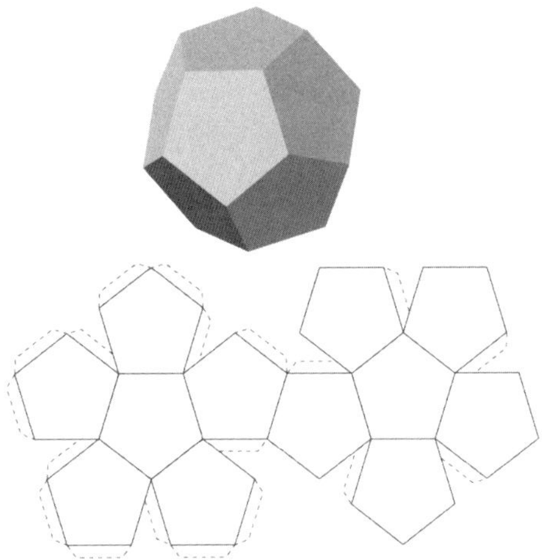

C7-14 Dodecahedron: Three-dimensional solid and two-dimensional net with tabs

LEARNING EXPERIENCE
Designing Product Packaging

- Have students investigate advertising and packaging using print or digital sources. Lead a classroom conversation that focuses on the ideas that students find in their research. Based on this sharing, have students develop a three-dimensional package (e.g., box, carton, container) for a product using the elements and principles of art/design. What do consumers need to know? How can the consumer be attracted to the product? What colour schemes need to be considered? What typeface and how many different fonts and sizes should be incorporated into the design? How much type/textual information and visual information needs to be included? Why?

- Direct students to design their package with non-objective design elements that flow around the surface of the form. When doing so, they should consider line, colour, texture, shape, positive and negative space, and at least one principle of design such as balance, repetition, or movement. An alternative or subsequent activity would be to have students incorporate found/drawn images into a packaging design. They could then compare the visual impact of non-objective versus objective design approaches (i.e., how does text and realistic imagery affect the viewer's focus/attention/bias when viewing a design?).
- Students can discuss how musical jingles and engaging visual logos/trademarks are both used in advertising. Research, view (in class or individually), and discuss examples from the contemporary and historical

C7-15 CD cover designs

media forms? Have students discuss this topic (i.e., the power of visual/ audio marketing campaigns) with various adults from different generations and then compare findings as a whole group in class.

Ethnomathematics

Ethnomathematics is the study of the relationship between mathematics and culture. The goal of ethnomathematics is to contribute both to the understanding of culture and the understanding of mathematics, and mainly to lead to an appreciation of the connections between the two. In this section, we will discuss some examples of designed objects that intersect both cultural and mathematical endeavours.

Cultural Circles

Circles are a part of peoples' understanding of the universe. Throughout history, humans have sat in storytelling circles, and they have used circles in their visual art, dance, music and musical instruments, in their cookware, and in their depictions of the shape of the hearth. Students find circles more meaningful when they see examples of circles as they occur in their world, from the concentric rings produced by dropping pebbles in water to the shape of the full moon in the night sky. When students come to realize that circles are a part of their everyday life, and that circles often characterize significant moments in their own sharing, learning and creating, this simple geometric shape can take on new meaning.

Euclid (300 BC) defined the circle as "a plane figure contained by one line such that all the straight lines falling upon it from one point among those lying in the figure are equal to one another" (Scott, 1974, p. 7). Vocabulary associated with the circle includes the centre (the point in the middle), the circumference (the length of the line forming the circle), the radius (distance from the centre to the circumference), and also related terms such as line segment, sector, chord, and tangent.

The word "circle" has been used in a broad range of contexts, beyond the mathematical shape. *Circle* can mean many things, such as a round enclosure or a number of people gathered in a ring. The circle can refer to any series ending as it begins, and perpetually repeated, like the cyclical or iterative patterns found in the measure of time, seasons, and life. The word can also denote a number of persons or things considered as bound together by some bond in a class, a set, a sphere of action or influence, such as a circle of friends, voices, companionship, and care. The circle can be broken, unbroken, within a square, or without a centre. People move, dance, sing, make music, and create visual art in circles.

Humans have come to understand the earth as spherical, rather than flat, and as one of the planets that moves in an elliptical orbit around the sun, rather than existing at the centre of the known universe. Knowledge about circles and curves (e.g., sphere, spiral, circle, helix, parabolic curve, elliptical orbits) has been applied to many areas of human understanding, creation, and expression.

LEARNING EXPERIENCE
Methods of Making Circles

- **Circle construction**: A compass is not always needed when one wishes to create a circle. Circles can be constructed by placing a pushpin into a sheet of paper on top of cardboard, and then attaching a pencil to a string and looping it around the pushpin point. Keep the pencil in an upright position and draw a circle by extending the string to its full length and then moving it carefully around the pushpin. Other methods include attaching a piece of wire, paper clip, or cardboard onto a pencil. Simply secure one point with a pin or another pencil, and rotate the end with the other pencil. Another method is by tracing templates of circles, such as coins, cans, and bottles.

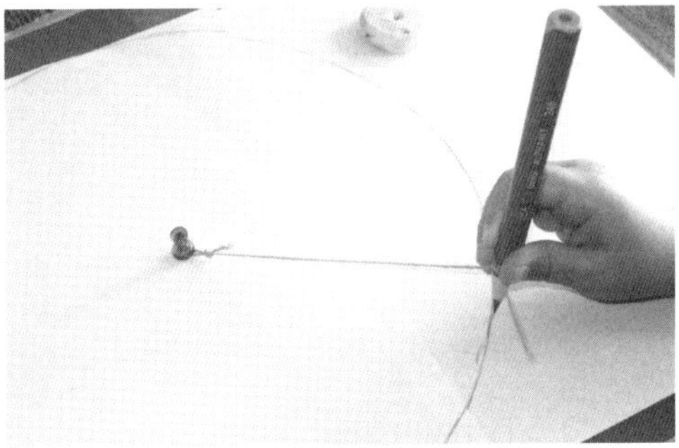

C7-16 Circle construction with string, pushpin, and pencil

C7-17 Ellipse construction with string, pushpins, and pencil

- **Ellipse construction**: A circle is a special case ellipse where the two foci are coincident (directly on top of one another; same point). To construct other ellipses, place two pushpins into a sheet of paper, tie off a piece of string, loop it around both pushpins (the foci), place the pencil in the string, and then trace the ellipse by keeping the pencil straight vertically and moving it around the pushpins slowly.

LEARNING EXPERIENCE
Measuring Distance with a Trundle Wheel

- Draw a circle with a 4.8 cm radius that is on cardboard. Use a protractor to divide the circumference into twelve equal arcs. Beginning with one point on the circle, number each point on the circle like a clock. Cut the circle out to form a wheel. Place a pin or pencil through the centre so the wheel will roll on a surface. Roll the wheel on a ruler. How far does the wheel roll in one complete turn? What is the length of each mark on the ruler? Use this wheel to make some measurements by rolling it on a surface (Johnson, 1974, p. 14). Use this technique to measure a length of rolled paper for a group mural.

LEARNING EXPERIENCE
The Potter's Wheel

The concept of circular motion is perhaps nowhere more visually stunning than the potter's wheel. Throwing clay on the potter's wheel can be exciting, but it is also difficult to learn how to do well. Experienced potters can make this activity appear so easy to do, as they create beautiful works of art and design using the technology of the foot-pedal or mechanically driven potter's wheel. In the words of Kemp (1996), "To one who has watched the potter at his wheel, it is plain that the potter's thumb, like the glass blower's blast of air, depends for its efficacy upon the physical properties of the medium on which he operates, which for the time being is essentially a fluid. The cup and the saucer . . . beautiful surfaces of equilibrium . . . formed slowly, under conditions of restraint which enhance or reveal their mathematical symmetry" (p. 27). Symmetry, or more specifically, the ability to recognize, control, and create symmetrical patterns in clay, plays a large part in creating successful pottery objects. Like glass-blowing, throwing clay on the potter's wheel involves a skillful combination of mathematical (e.g., symmetry, timing, applied force) and aesthetic (e.g., design, elements/principles) awareness and technique.

- Have students watch videos of clay objects being formed on a potter's wheel, and/or invite an experienced potter to conduct a demonstration for students. Encourage participation if this is possible.

C7-18 Hand-crafted Thai Celadon pottery tea set

Mandala and Sacred Circles

Mandala is the Sanskrit word for "circle." Mandala shapes are found in nature in the plant and animal kingdoms, as well as in constructed objects, in art, and in architecture. The mandala has been interpreted in some cultures as a symbolic replica of the world, a geometric projection of pattern wherein the axis of the mandala represents a line of communication between spiritual/metaphysical powers above and humanity below. The Hindus used mandalas for meditation and mandalas appear in many other cultures and religions (e.g., Aztec, Maori, North American Aboriginal, Navaho, Buddhism, Christian Gothic).

The earliest mandalas were ancient sun wheel designs that were scratched or carved into rock surfaces. The yogis and priests of early Hinduism and Buddhism marked circles around themselves as representations of their sacred space. The circle was equated to the circular horizon surrounding them. The circular mandala form was also translated as the letter "O," a critical vowel sound in Hebrew and Greek, imitating the breathing needed to produce the correct sound. It also resembled the organ of sight, the human eye. This biaxially symmetrical letter was the root of the words *optic*, *oral*, *ode*, *order*, and *opinion* in English, *ops*, *orion* (boundary), and *organon* (musical instrument) in Greek, and *oeil* in French. In Spanish, *orillo* means circumference. The Latin *orbis* was anything round, a ring or a representation of power on earth, while *ovum* was an egg. "O" was also the Egyptian hieroglyph for horizon and was used for words meaning "whole." Carl Jung, a psychologist, encouraged his patients to paint mandalas as a tool for self-discovery.

The ring-shaped mandala is broken up by two sets of intersecting lines resulting in a representation of the four corners of the earth. The number "four" is sacred among the Blackfoot people specifically, and the plains First Nations more generally. During the sun dance, the most sacred of the Blackfoot ceremonies, the entire camp moved four times. On each of the four days of the ceremony, certain prescribed activities were carried out. Ritual movements are repeated four times. Prayers were made from each of the cardinal directions.

Other related areas involving circles and cultural practices include sacred circles, "dinergy" (term coined by Doczi [1981] meaning the pattern-forming process of the union of opposites), and the creative energy of organic growth.

LEARNING EXPERIENCE
Creating a Mandala

- Research various types and styles of mandala circles, and then have students design their own mandala pattern with geometric shapes, colours, and/or images that have personal meaning to them. Consider the elements and principles of art/design, especially balance, symmetry, positive shapes, and negative spaces.

C7-19 Student mandala designs
(colour plate xxx)

Cultural Research

Every culture has designed objects for functional and aesthetic purposes. Most of these objects incorporate both mathematical and visual art concepts. Examples for research include clothing, masks, headgear, cloth, weaving, totems, symbols, pottery, North American pioneer quilts and patchwork, origami, and so forth. The Ukrainian Easter egg learning experience that follows is one such suggestion.

LEARNING EXPERIENCE
Drawing an Accurate Easter Egg

1. Draw a circle with centre O, then draw the diameter AB through point O.
2. Construct the perpendicular bisector (i.e., a line at right angles to the diameter that divides it exactly in half) CD of line AB, and extend it beyond the circle at the top. To construct the bisector, set the compass at a length between the radius and diameter (greater than OB, but smaller than AB). With B as the centre, cut an arc above and below. With A as the centre and same length, repeat this step. Draw a straight line CD through the two points where the two arcs intersect at the top and at the bottom.
3. With A as the centre and radius AB, construct arc BE. With B as the centre and the same radius length, construct arc AE.
4. Draw a straight line AF through point D on the circle (i.e., where the perpendicular bisector intersects the circle).
5. With D as the centre and DF as the radius, now construct an arc GF to finish the egg contour.

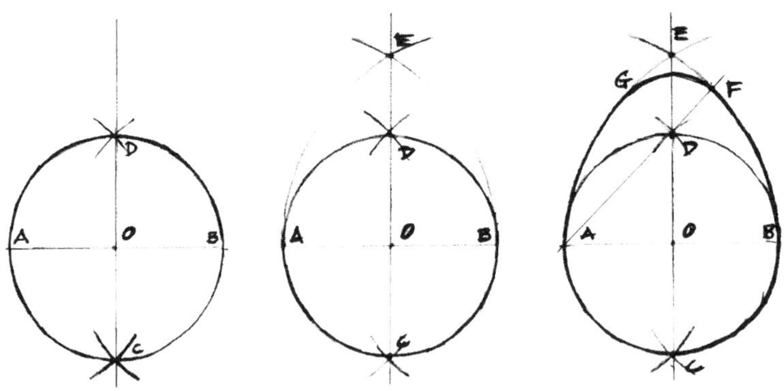

C7-20 Egg construction using a pencil, compass, and straight edge (ruler)

6. Decorate your "egg" using design decisions based on the elements and principles of art/design learned previously in the course.

C7-21 Intricate egg painting designs using wax resist and dyes (colour plate xxx)

LEARNING EXPERIENCE
Currency Design

- Research various cultures and the currency (i.e., printed, coin, or other forms) used in their everyday financial interactions. Observe the images and symbols used in the design of their currency, as well as the elements and principles of art/design.
- Create a group or a shared class currency (money) system and related images/designs. Consider the elements and principles of art/design in the development of this new currency.
- As a possible extension, connecting to issues in geography, economics, and social science, research and discuss what advantages and disadvantages different forms of national/regional/local currencies may have. For example, what implications did the establishment of the Euro have in continental Europe? Why might some countries choose not to take part in this shared currency (benefits/risks)? What effects could a significant downturn in a country's economic status (e.g., Greece in 2011) have on that country's administration and people, as well as on other countries sharing the currency? What about an international carbon credit system to help preserve the environment? What factors might affect such a major change in "global currency" (reasons for and against adoption)?

Time

Time is an important factor for humans in most cultures. For example, from ancient times, seasonal celebrations have been a centre of community life. Midsummer, or the summer solstice, when people in the northern hemisphere enjoy the longest day and shortest night of the year, is a time of hope and renewal celebrated in many northern nations. The Millennium Clock Tower, constructed in the year 2000 in Calgary, Alberta, to mark the arrival of the new millennium, has become a focus for community celebrations, bringing local residents together each June 21 to whitewash the graffiti wall, symbolically cleansing and renewing the shared canvas for the coming year.

People first measured time by the rising of the sun and the phases of the moon. Babylonians and Egyptians have been recognized for their developments of various types of calendars, which are written documents designed to measure and track the progression of time. Clocks measure shorter periods of time. One of the first instruments used to show the time of day, but not night, was the sundial. Hourglasses filled with sand or water were also invented to measure a given amount of time. Present units of time are defined in terms of the movement of the earth with reference to the stars, closely related to the science of astronomy (Johnson & Glenn, 1961, p. 20).

"We use time as a form of measurement and order—the calendar, the season, days of the week, holidays" (Naested, 2010, p. 234). In ancient history, the Babylonians thought there were 360 days in a year. They "drew an angle that was 1/360 of a circle. This was their unit angle. It is now known that there is approximately 365.25 days in a year, but the Babylonian unit angle is still used as the basis for measuring angles or turns in a circle. The unit angle has a measure of one degree" (Hoffer, 1977, p. 288).

LEARNING EXPERIENCES

Calendar Design

There are many forms of calendars that various cultures and religions have developed throughout history. How do calendars relate to art and mathematics? For mathematics, numbers, of course, form a significant part of the design of various kinds of calendars (e.g., someone had to calculate the rotation of the sun, growing seasons, length of shadows, and size of the moon (monthly waxing and waning). The artist was employed to elaborate on the mathematicians' numbers to make concepts understandable to the "common folk," but also to elaborate and embellish the ideas for aesthetic reasons.

- Research various calendars from different countries and cultures. Observe the different images and symbols that are used. Assign each student to a month of the year (possibly the month of their birthday). Have students develop a calendar for their month that features appropriate images for the month and time of year. Consider the elements and principles of art/design.

LEARNING EXPERIENCE

Time and Tide Artwork

A famous line by Chaucer says, "Time and tide wait for no man." Coastal tides are linked to the moon's orbit around the earth and the related gravitational pull. Measuring the moon phases has been a fascination for humans throughout history. "The phases of the moon make a complete cycle from new moon to full moon and back again every twenty-eight days. The year is three hundred and sixty-five days long—roughly" (Stewart, 1995, p. 3).

Strange and beautiful sky events have regularly occurred involving the sun and moon. For example, a rare full-moon lunar eclipse during the winter solstice of the moon's cycle occurred on the night of December 21, 2010. This event had not occurred for 372 years, according to contemporary astronomers. One of the authors watched this spectacular event with his daughter, bundled in blankets on lawn chairs with hot chocolate—the strange magenta moon disappearing and reappearing from about 3:00 to 4:00 a.m. Eastern Standard Time. It was a mathematical, time-dependent, aesthetic human experience not soon to be forgotten. Have students research this eclipse on the Internet to watch the public video documentary archive of this rare natural phenomenon that must have amazed (and perhaps frightened) humans in centuries past.

- Have students research the lunar cycles (waxing/waning, summer/winter solstices) that occur regularly in their hemisphere. Also have them research the dates of unique events, such as lunar and solar eclipses, and how people have responded to these events historically. Further, have students explore how the earth's tides are tied to the lunar cycle. Students can then develop a work of art (drawing, painting, sculpture, video) that relates to these various phenomena.

Sundials and Clocks

Teaching young children how to tell time (both analog and digital, although the former is arguably much more difficult) is among the most challenging topics in the early years mathematics curriculum. What may appear simple to adults is actually a highly complex set of related concepts and skills that a student must master in order to fully understand the telling of time. For example, the analog clock measure involves a base-60 (sexagesimal) system for seconds and minutes, a base-12 (duodecimal or dozenal) system for hours, a base-24 system for military/airport time, and base-4 (quaternary) and base-2 (binary) systems for references made to quarter to/past and half-past, respectively. Complicating matters, most students now rarely encounter analog clocks in their daily lives, where time is usually represented in a digital format.

C7-22 Big Ben Clock Tower of British Parliament in London, UK (colour plate xxx)

Sundials, clocks, and watches help us to gain a clearer understanding of our sun and moon, and thereby help us to better understand, plan for, enjoy, and remember our lives. A sundial is an instrument to show the time of day by the shadow of a gnomon on a horizontal plate, or on a cylindrical surface. The sundial is one of humankind's earliest achievements—a mechanism designed to mark the passage of time. The ancient Egyptians used sundials. A sundial is mentioned in the Old Testament (Isaiah 38:8) and would have existed around 1700 BC. Berossua, a Chaldean (ancient Babylonian), wrote an expert treatise on sundials around 1300 BC that was the definitive work on sundials for over a millennium, until the Christian era. The Greeks and Romans used sundials extensively. In fact, the indicator on the sundial that casts the shadow is still called by the Greek word "gnomon," which means "pointer."

C7-23 Large clock inside the Musée D'Orsay (formerly a train station) in Paris, France

The gnomon, the part that casts a shadow on the

base of the horizontal sundial, indicates the time of day. The angle must equal the latitude of the location where the sundial will be used. The latitude of Calgary, Alberta, for example, is 51°03´ N (51 degrees, 3 minutes, north); the angle between the base and the gnomon would therefore be 51°03´ degrees.

The Arabians were renowned for more complicated sundials of greater accuracy and scope of design. Elsewhere on the globe, the Chinese were great users of sundials and the Aztec and Inca civilizations in the Americas were also familiar with these devices. The development and use of the sundial was, and remains, a global endeavour.

Time represented by a sundial is referred to as Local Apparent Time (LAT). It is based strictly on the apparent motion of the sun, so that when the sun is viewed to be directly over the observer's meridian, the LAT would be noon. By the eighteenth century, clocks and watches began to supersede sundials. These machines had one significant advantage—they were not subject to the unpredictability of clear skies. Ironically, however, clocks/watches were often unreliable, particularly in their early development, and so people needed to depend on sundials to reset the "true" time.

With the development of accurate analog and digital watches, the need for sundials has diminished. Nevertheless, sundials should not be forgotten. They are still one of the most reliable forms of timepieces available.

LEARNING EXPERIENCES

Creating Sundials and Model Clock Towers

- Have students research and then create their own sundial for telling time outdoors.
- Have students develop a three-dimensional scale model of a clock tower using cardboard or foam core. Consider size, shape, and other elements of art/design.

C7-24 Sundials allow the viewer to tell time using the sun's rays and cast shadows

Games and Films

Cultures have also developed competitive, cooperative, and individual games throughout history. Such games assist students in the development of critical thinking and visualizing abilities in both visual art and mathematics through activities such as matching, describing, identifying, classifying, measuring, and comparing. A short list of such popular games might include:

Battleship
Billiards
Checkers
Chess
Four-square
Go
Hopscotch
Mancala
Sports (baseball, football, basketball)
SET: The Family Game of Visual Perception

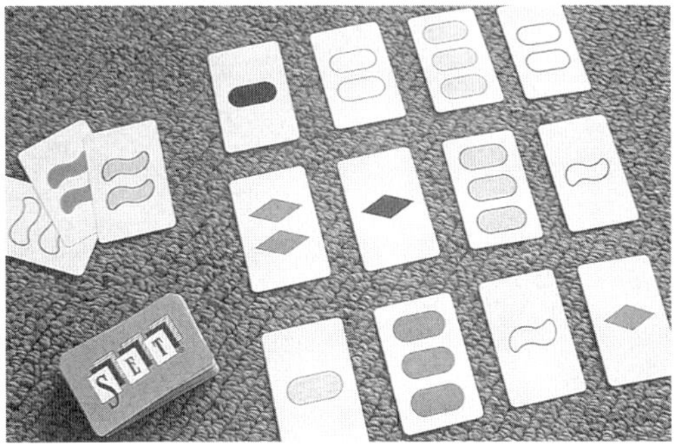

C7-25 SET: The Family Game of Visual Perception, created by population geneticist Marsha Jean Falco in 1990, requires both visual and numeric (math and art) recognition/thinking.

LEARNING EXPERIENCE
Game Design and Observation

- In pairs or small groups, have students create and design their own game that involves the reinforcement of selected concepts/skills from both mathematics and visual arts. When completed, have groups present their games to their peers, and then have students observe peers playing their games in order to ascertain if, or in what ways, the learning goals/objectives of the game are achieved. Alternately, or in addition, have students design similar games for a younger audience, allowing them to watch younger children play their game and observing how the mathematical/art concepts are reinforced. Allow students to debrief after observations, modify their games accordingly, and then observe actual gameplay again, repeating the reflective cycle.

Films Involving Connections Between Mathematics and Visual Arts

In 1959, Walt Disney produced a film for the U.S. government to increase student interest in mathematics in light of the Cold War and "race for space" that was taking place at that time. *Donald Duck in Mathmagic Land* contains a series of mini-chapters that each showcase the connections between mathematics and some other field/theme. Mathematics is discussed in terms of ancient history (Greeks), music, visual arts, games, science/nature, and technology/architecture. A similar, more recent (2006) series, *Life by the Numbers*, hosted by Danny Glover, once again provides students with fascinating examples of mathematics connections to various areas of everyday life and culture (e.g., computer-generated imagery (CGI), forensic science, nature patterns, probability theory, and technology). Also recommended in terms of video resources which help students better appreciate the connections between visual art and mathematics are *The Story of 1* (2005, with narrator Terry Jones), *Flatland: The Movie* (2007, based on Ed Abbott's classic Victorian novel), and *The Story of Math* (2010, narrated by Oxford math professor Marcus de Sautoy).

LEARNING EXPERIENCE
Video Documentary Creation

- In pairs or small groups, have students create their own short video documentary of some topic they feel contains elements of both mathematics and visual art. Students should draft the narrative, storyboard the sequences, act in and film the shots, edit the final footage with titling, and then present it to the class for discussion and critique. All of the class videos can then be uploaded (with appropriate permissions) to a video sharing server (e.g., Vimeo, YouTube) as a class "channel" to be shared with friends, family, other classes/schools, and, by virtue of the web, the world at large.

Summary

In Chapter 7, we have discussed the connection between mathematics and designed objects. These include architectural details and decorations, two- and three-dimensional shapes, cultural and sacred circles, calendars, time-related objects, currency, games, and film. We hope that the reader/learner/teacher will be able to incorporate the ideas presented in this chapter to develop learning experiences which encourage critical thinking and visualizing abilities in both visual art and in mathematics. In so doing, the curriculum can be increasingly examined through both artistic and mathematical lenses, leading to a raised awareness among both teachers and students.

8 Making Connections

Teaching and Learning Between the Lines

> *I do think many children can become interested in mathematics—particularly geometry [through art]. One of the modes of entry into both music and the visual arts is mathematics—at the same time there can be a kind of reversal of this, in which the arts become an entry into mathematics.* (Alex Colville, 2003, personal communication)

As we have highlighted throughout this text, mathematics and visual arts share many commonalities throughout history and within our contemporary world. Furthermore, we have attempted to demonstrate through many examples how teaching and learning strategies from both of these disciplines can be combined, or integrated, to enhance and enrich the learning experiences of students—hence the idea of "teaching and learning between the lines."

When speaking of integration, or interdisciplinary learning, it is first helpful to understand what this pedagogical approach stands in contrast to. In today's schooling, we have inherited the organizational structure of the various disciplines, which are arbitrary groupings of related knowledge and skills. A key question then becomes, Should the disciplines exist as curricular organizers in contemporary schooling? Should they be abandoned, or at least modified, in light of a new or different set of required "basic skills" in our post-modern society and workplace? There are obviously many opinions on these significant questions. What follows is a brief description of both sides of the main argument: Is the discipline structure in today's schooling a balm or a bane?

The Balm of the Disciplines

Vindicators of a discipline-based curriculum are myriad. Its longstanding existence as a systematic form of educational organization speaks to its utility. Mason (1996) states:

> Scholars have historically organized and taught human knowledge in the context of the major disciplines of mathematics, philosophy, the sciences, the arts, and the humanities and the subject matter areas that have been built around them (e.g., social studies). Many consider the academic disciplines a useful framework for organizing curricular content because they constitute the most sophisticated ways yet developed for thinking about and investigating issues that have long fascinated and perplexed thoughtful individuals . . . they become, when used relevantly, our keenest lenses on the world. . . . Introducing students to the forms of discourse and inquiry associated with the disciplines and the subject matter areas can provide them with valuable mechanisms for making sense out of a complex world. (p. 263)

Interestingly, in a similar in-depth analysis, Case (1994) notes that the disciplines, in their very structure and essence, bear a form of integration themselves. He states that "the disciplines, for all of their narrowness, provide integrative principles—disciplines are fields of inquiry that share standards of evidence, fundamental explanatory concepts, and methodological procedures" (p. 86). Gardner (2000) explained the history and nature of the disciplines at length. As the creator of Multiple Intelligences Theory, which has been widely used by integration theorists, Gardner is somewhat ironically a devout advocate of a discipline-based curriculum:

> Disciplines have endured over the ages, even as their identities and boundaries have shifted. At any given moment, the disciplines represent the most well-honed efforts of human beings to approach questions and concerns of importance in a systematic and reliable way. (p. 144)

The Bane of the Disciplines

Many proponents of an integrated curriculum have rallied bitterly against the discipline-based model. In the rationale of their integration document, Attenborough, Pattison, Patsiatzis, and Muller explain:

> In the last decade, the education system at the secondary level has attempted to create students who would be experts in specific fields of study through the concept of delivering strictly specialized courses and in some cases creating specialized school buildings. This resulted in overspecialization where students acquired "knowledge" as disjointed bits of information. (1997, p. 2)

Mason claims that "The traditional curriculum fails to meet the needs of students in a complex, technologically advanced, interdependent world" (1996, p. 265). This concern is reiterated by Schramm (1997):

> The traditional subject-centered curriculum does not reflect the social reality that our students experience in their everyday lives. According to

> Jacobs (1991), the discipline-oriented structure of traditional curricula has contributed to the current crisis in American schools. Many young people are physically dropping out, while others drop out mentally, sitting passively in their classrooms waiting to survive the day. (p. 5)

Fogarty (1991), in describing her first of ten types of integrated models, criticizes the fragmentation that she believes plagues the discipline approach to education: "In middle and secondary schools, the disciplines are taught by different teachers in different locations, with students moving from room to room. Each separate encounter has a distinct cellular organization, leaving students with a fragmented view of the curriculum" (p. 62). An even more severe denouncement of the discipline model comes from Brady:

> The academic disciplines began as means, and they have become ends. Now, in most classrooms, the reality the disciplines were designed to illuminate is of less consequence than the history, vocabulary, methodology, and procedures of the disciplines themselves. . . . Despite their nearly universal use as organizers of education, the traditional disciplines are not the best available tools for teaching about reality. If we looked at them critically, we would see that they are poor material from which to build a general education curriculum. They ignore vast and important areas of knowledge. They give instruction no overarching aim. They have nothing to say about the relative importance of various kinds of knowledge. They do not give students a mental framework for organizing and relating what they are taught. Because they cannot be made to integrate with one another, they fail to disclose the systemic nature of reality. . . . In fact, "discipline-based general education" is an oxymoron. (1993, p. 439)

Naested's textbook, *Art in the Classroom: An Integrated Approach to Teaching Art in Canadian Elementary and Middle Schools* (2010), is dedicated to an integrated approach involving art and other subjects. Reasons presented for art/subject integration include that integration

- Stimulates images and enhances student learning by involving students visually, physically, and emotionally;
- Connects course content, concepts, and skills;
- Addresses multiple intelligences or learning styles of students;
- Develops alternative avenues for expression that are not limited to words and numbers;
- Connects the art class with the outside world: television, advertising, dress/costume, architecture, city-planning, the natural environment, and other cultures; and
- Opens alternative approaches for learning, teaching, and communication (2010, pp. 4–5).

If integration is to be adopted by teachers, how should it be defined and implemented? In this next section, we will present some caveats, or warnings, for teachers to consider when planning for integrated learning experiences; some examples of various integrated learning models; and, based on a research study by one of the authors (Jarvis, 2001), a description of qualities germane to integrated learning, as well as four key elements when planning rich, integrated learning experiences.

Integration Caveats

Both mathematics and visual arts educators have expressed valid concerns regarding the defense of their discipline's integrity within interdisciplinary curricula. For example, Jarvis (2001) recalls a university mathematics education professor who once shared his deep concern regarding the ridiculously simplistic mathematics skills required for what he described as nefariously integrated projects at the secondary level. A (negative) reciprocal anxiety also exists among art educators. Eisner (1999) presents three categories of arts outcomes: arts-based, arts-related, and ancillary. It is to the latter type that he relegates the use of art in boosting academic achievement in other disciplines. Although he does not oppose an integrated curriculum, Eisner strongly maintains that

> We do the arts no service when we try to make their case by touting their contributions to other fields. When such contributions become priorities the arts become handmaidens to ends that are not distinctively artistic and in the process undermine the value of art's unique contributions to the education of the young. (p. 149)

Other visual arts specialists have articulated similar concerns. Cormack (1991) sardonically explains, "Cranked-out rabbits and turkeys are not art education, listening to records as a time killer is not music education, and being a pilgrim does not a Thespian make" (p. 42). Grauer also notes,

> Illustrating a story, the title page for a social project, colouring in the geometric shapes after a math lesson, are all examples of this type of rather inane "art" experience. . . . There is no substantive art learning taking place, rather art is used as a motivator at best and a time filler at worst. Again, we come back to the need to understand the discipline of art before an integrated experience can have educational goals. (1991, p. 26)

A second area of caution is the various problems associated with the implementation of integrated models. For some educators, integration has simply meant licence for disorganization and a smokescreen for lack of planning. Marshall provides this disclaimer: "In the chaotic conditions created by misunderstood 'individualism' and 'unlimited freedom,' integration of anything can change rapidly into fragmentation of everything" (1970, p. 14). Other poignant issues that arise are the following: lack of planning time, forced connections regarding themes, lack of teacher expertise, training, and approval, student disinterest during extended projects, lack of relevant materials, and the abandonment of discrete instruction in basic skills (Barth, 1995; Case, 1994; Hanley, 1994; Woodbury, 1998). In light of these multiple barriers and caveats, Mason's (1996) conclusion seems pertinent:

> The works of artists, scientists, and other creative individuals who have advanced knowledge and culture by transcending the boundaries of the disciplines attest to the value of striving for an integrated curriculum. The efforts of those who have examined the practice of curriculum integration over the last decade can provide a foundation for educators seeking to establish interdisciplinary programs in schools. Only by proceeding with caution, however, and by attending to the potential obstacles identified here, can educators achieve an integrated curriculum that will result in a better curriculum. (p. 271)

When considering integrating art with other subjects, Naested (2010) realizes that

> Art educators may question how art concepts, values, skills, and content can be integrated with other subjects while still maintaining the inherent values and integrity of art education curriculum. Watering down the art program or presenting art concepts inadequately in an integrated system is considered invalid, particularly when integration is not clearly understood and not properly carried out. (p. 28)

Eisner (1991) was not only an early proponent of arts integration, but he also realized potential difficulties or shortcomings:

> The integration of one subject with others is not without potential costs. When integration is aimed at, the special features of a subject are often compromised in the process . . . Yet I do recognize that any approach has its trade-offs. I value integration, coherence and unity. I would urge that connections between the arts and other subjects be made when they can, as long as the values of art are not diminished in the process. (p. 19)

Models of Implementation

As Irwin and Reynolds note, "A first step on the path to clarification is for educators and academics to recognize that there is no consensus about the meaning of integration" (1995, p. 15). Germane to this elusive definition is the plethora of integration models that have been proposed. They have included a survival-based model (Brady, 1993), an ecology-based model (Jardine, 1990), a problem-based model (Pring cited in Irwin & Reynolds, 1995), and a theme-based model (Jacobs, 1989). In more general terms, theorists such as Fogarty (1991), Badley (cited in Irwin & Reynolds, 1995), and Woodbury (1998) have presented categorical analyses of a variety of models, often in a continuum format. Bickley-Green (1995), in her paper "Math and Art Curriculum Integration: A Post-Modern Foundation," concludes with the following statement regarding math and art integration:

> The study of integrated math and art in curricula shows potential for congruent activities that will enhance learning in both disciplines. Educators can do further research to demonstrate how art and geometry are cognate cognitive structures; teachers of math and art can work together to identify and instruct developmentally appropriate and integrated math and art activities for learners; and curriculum writers can reform the two courses of study by considering the related webs of meaning between the disciplines before creating impenetrable boundaries. (p. 17)

Curriculum Integration Research Study Highlights

In 2001, Jarvis conducted a graduate research study involving the integration of mathematics and visual arts in secondary school curricula. Participating students completed several extended, integrated projects and were subsequently interviewed regarding their learning experiences. Findings from the research study led to a list of perceived qualities of integrated learning (complexity, playfulness, universal appeal) and also to the highlighting of four key characteristics of integrated curriculum planning (creative space, adequate time, legitimate matter, infectious energy). These perceived qualities and characteristics are summarized on the pages that follow. Selected sample projects from the study (Golden Proportion, Calculus) can be found in Appendices C and D.

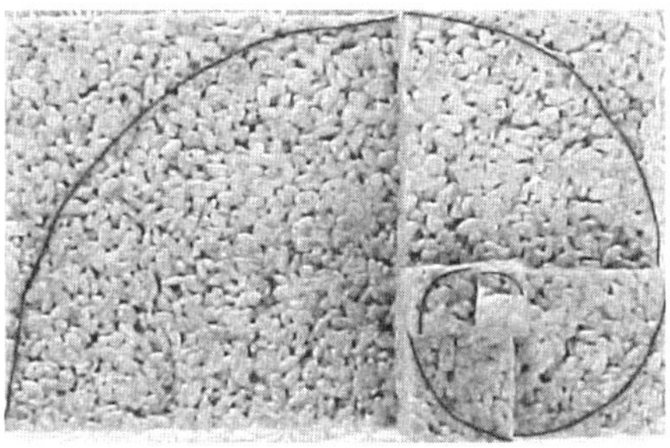

C8-1A Golden Rice Krispies square tower (colour plate xxxi)

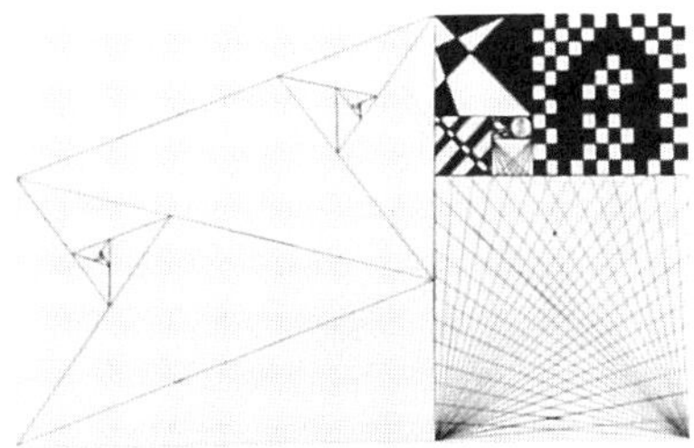

C8-1B Golden "SMARTWORK" design spelling the title with letters in each diminishing square

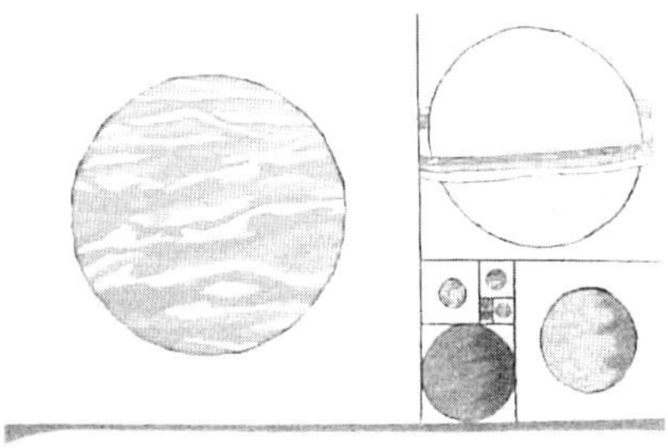

C8-1C Planets arranged by descending sizes of circumference (colour plate xxxi)

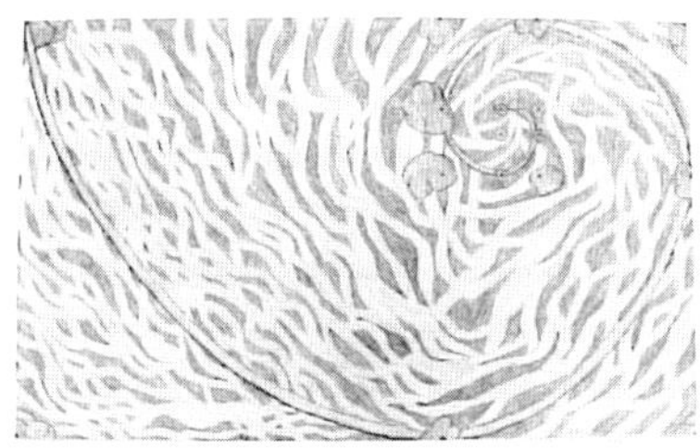

C8-1D Golden spaghetti noodles, meatballs, and fork tines (colour plate xxxi)

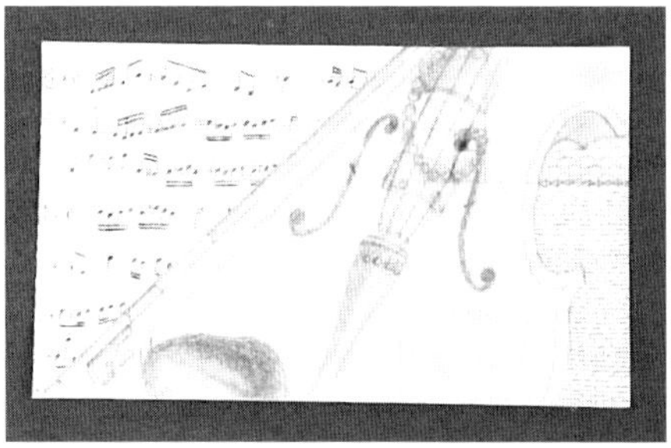

C8-1E Golden music and string theory theme

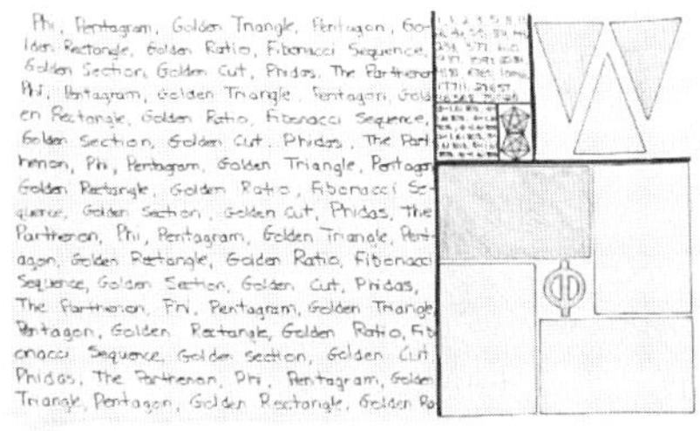

C8-1F Golden shapes and vocabulary
(colour plate xxxi)

Qualities of Integrated Learning Experiences

Complexity: Sylwester (1995) hypothesizes that a brain-based curriculum of the future might "increase the importance of such subjects as the arts and humanities, which expand and integrate complex environmental stimuli, and reduce the importance of basic skills and forms of evaluation that merely compress complexity" (p. 23). While the integrated assignments were designed to reinforce expectations from the curriculum, they did so in a way that required students to both process new information through a variety of senses, and then creatively apply the new learning through complex expressions of communicated understandings. This non-traditional form of mathematics learning is not only unfamiliar to students, but can also be terrifying for the educator. Wiggins and McTighe (1998) explain this phenomenon:

> As teachers, we are trained to believe our job is to remove doubts and explain things. Teaching is meant to take away the dead-ends, false starts, and surprises of inefficient inquiry. But teaching engagingly and effectively for understanding often requires that we persist in asking questions, delaying or avoiding giving answers, confronting students with problems, and putting mysteries and the need to rethink things constantly before them. (pp. 141–142)

When asked during interviews if a math/art course based entirely on integrated units of study would be a good idea, many students supported such a concept. With regard to interdisciplinary educational complexity, the following student verbalized a common attitude that surfaced several times and in various forms within the interviews and questionnaires: "I think that it would probably enrich the learning as opposed to distracting from it, because school is combining everything you learn and applying it to the real world. You take things from one course and stuff from another course and hook them together." Conspicuously evident within the fabric of the projects themselves, and permeating the processes of comprehension, creation, and communication that each student underwent, complexity was recognized as one of the unique qualities germane to the integrated learning process.

Playfulness: Another quality that surfaced through daily observations and the research instruments was that of playfulness, or the sense of fun or festivity experienced by students within the integrated learning. Pedoe (1976) refers to this quality as what he calls the pleasure principle, and encourages practitioners to "consider this principle of fundamental importance in all education, and especially in mathematical education" (p. 13). Davis and Rimm (1994) describe the elements of humour and playfulness that figure into the character traits of the creatively gifted student:

> An especially frequent trait of creative students is a good sense of humor. Humor is first cousin to the ability to take a fresh, childlike, and playful approach to problems. Many discoveries, inventions, and artistic creations are the result of "fooling around" with ideas and playing with possibilities, unconstrained by well-learned habits, traditions, and conformity pressures. Both Carl Rogers and Sigmund Freud agreed that regression to a more childlike, fanciful, playful state of mind is an important feature of creative thinkers. (p. 35)

Some of the more insignificant activities involved in the assignments, such as the measuring of the proportions of student volunteers (i.e., full height vs. height to navel) in front of the class, the lesson on "how to singe one's Renaissance drawing without lighting it on fire," or the hands-on template construction as an introduction to tessellations were, ironically, the types of moments that students indicated were most memorable. They certainly featured elements of playfulness and were only possible within a learning environment where students felt safe enough to explore, laugh, and admit that learning can be fun. What follows are three student quotations along these lines:

> I realized that it could be exciting and fun to play with the concepts, and manipulate them however you saw fit. I enjoyed this project because it caught my interest, and if more people get excited about the project, as I did, it could be beneficial to do every year.

> It's interesting learning where math fits in to everyday life. It's also neat to see the other parts of math that you don't usually see in class.

> I liked how the [projects] brought two things together—it was fun and you were learning too.

The notion of playfulness ties directly to student motivation. Gardner (2000) notes the connections between fun, pleasure, play, and motivation in the following:

> Flying in the face of the behaviourists, who tied motivation directly to the receipt of tangible rewards, researchers now believe that learners are best served when their motivation is intrinsic: when they pursue learning because it is fun or rewarding in itself, rather than because someone has promised them some material benefit. (p. 76)

Therefore, playfulness, as encouraged through the assignments and as found to be a motivator for both student and teacher alike, forms a second quality germane to the integrated learning process.

Universal appeal: A third ostensible quality of the integrated learning context is that of universal appeal. The questionnaires, interviews, and parent/guardian response forms all suggest that the vast majority of students enjoyed the projects. Different students liked the integrated learning for different reasons. For the gifted, or if not formally recognized as such, the brightest students, these projects provided an ambiguous space in which they could explore and express themselves within flexible parameters. Davis and Rimm (1994) describe the common learning styles of these types of students:

> Gifted students tend to be independent self-motivated learners more than teacher-motivated. They need and enjoy learning tasks that are unstructured and flexible, rather than the highly structured tasks needed by less able students. They prefer active participant approaches to learning, rather than spectator approaches; and they have "well-integrated perceptual strengths," meaning that they can learn through varied sensory channels, including auditory, visual, tactile, and kinesthetic. (p. 32)

For the average student, the projects offered alternative means for demonstrating their understanding through forms of communication and creative application that may not have been otherwise available through more traditional assessment methods. The scope of the assignments was broad enough and the time frame long enough that students of all levels of ability could often capitalize on their personal assets through their selection of integrated project ideas, and minimize their liabilities via cooperative group work, parental and teacher support, and personal research. In a related vein, Gardner (2000) notes that his theory of multiple intelligences ties motivation to existing talent: "The theory of multiple intelligences suggests another factor: people may be most motivated to learn when they undertake activities for which they have some talent. In pursuing such activities they are likely to make progress and avoid undue frustration" (pp. 76–77).

Successful integrated learning experiences strike an effective balance between the reinforcement of curricular expectations and the release of the individual's creativity and imagination. This balance may very well account for the qualities of complexity, playfulness, and universal appeal that integrated learning experiences apparently hold for students of different gender, ages, and academic abilities.

Characteristics of Integrated Curriculum Writing

Newton, in his imaginative explosion of ideas regarding the physical universe, was intimately familiar with the four modern elements of space, time, matter, and energy. Nearly three centuries later, Einstein paired this fundamental quartet with his theories of space-time continuum and energy-mass equivalency. As a tribute to both of these inspirational life-long learners, the four perceived characteristics of integrated curriculum writing are hung upon these historical posts.

Creative space: The integrated project must be designed in such a way as to provide students with sufficient space to creatively personalize the projects involved, according to the limits of each individual's imagination. This has been referred to elsewhere as the implementation of "parametric creativity" (Jarvis, 2009), or creativity within prescribed boundaries. In Complexity Theory, a similar concept is presented in the form of enabling constraints—rules or conditions that often lead to creative ideas and actions.

Adequate time: The extended integrated project must allow for an adequate amount of time in which students may consciously and subconsciously develop and wrestle with ideas. This struggle must transpire over a period of sufficient time in order for students' work to fully emerge into its mature form. By setting tentative due dates and allowing approximately one month per project, this invaluable time was procured for students. From the teacher's standpoint, this month-long time period, with carefully planned and placed group work sessions, computer lab investigations, art lessons, and teacher/student conferences, was deemed appropriate.

Legitimate matter: Above all else, the integrated project must feature meaningful questions or activities that tie directly to, and serve to reinforce, the related curriculum from both or all disciplines involved. The knee-jerk reaction to most integrated educational endeavours involving art is to suspect a lack of legitimate matter and to relegate the learning experience to a meaningless charade designed to increase motivation, fill time, or merely entertain students. Accusations of this sort are simply not possible to substantiate if the integrated learning experience is conscientiously tied to both the mathematics and the visual arts curricula. Regardless of which disciplines an integrated curriculum seeks to connect, it must strive to do so in such a way that the related curricular expectations are clear and justifiable. Only in this manner can increased student motivation be genuinely construed as legitimate.

Infectious energy: The integrated project is most successful when powered by three sources of infectious energy: that of the preceding students, that of the teacher, and that of the student participants. As any art instructor who has kept, over the years, an ever-expanding portfolio of student work will know, the learner likes to see examples. A variety of examples that demonstrate a range of ideas, ability and success is best. Some students argued that as many examples as possible would be appropriate to stimulate ideas. Others claimed that no, or relatively few, examples should be shown, as they tend to hamper creativity and actually retard progress. The ideal strategy likely lies somewhere in between, showing a few good exemplars to all students and then having many other samples ready to privately show the students who may require more guidance. The second form of infectious energy comes from the instructor. The introduction of integrated projects is enhanced when teachers adopt a variety of available methods and media (e.g., videos, books, computer software, posters, student work samples, and role play). Ongoing enthusiasm regarding the new learning and integrated projects can be maintained by teacher-student interviews or assessment checkpoints. Final presentations of student-created work is always exciting.

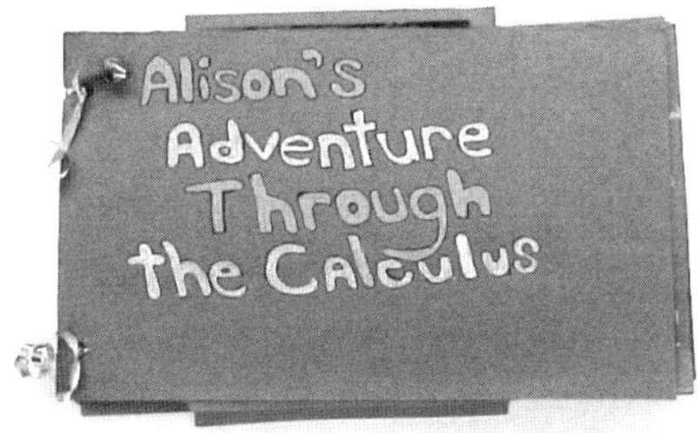

C8-2A Calculus project: Alison's "Adventure through the Calculus" illustrated storybook (colour plate xxxii)

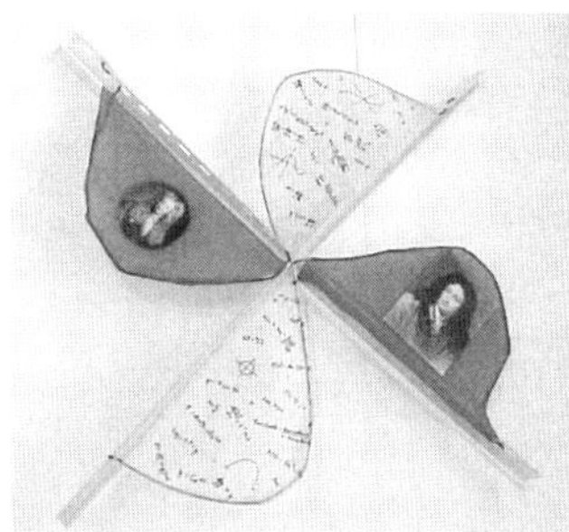

C8-2B Calculus project: Mobile featuring two creators of the Calculus (Newton & Leibniz) (colour plate xxxii)

C8-2C Calculus project: Newton poster illustrating his quotation about finding pebbles on the seashore (colour plate xxxii)

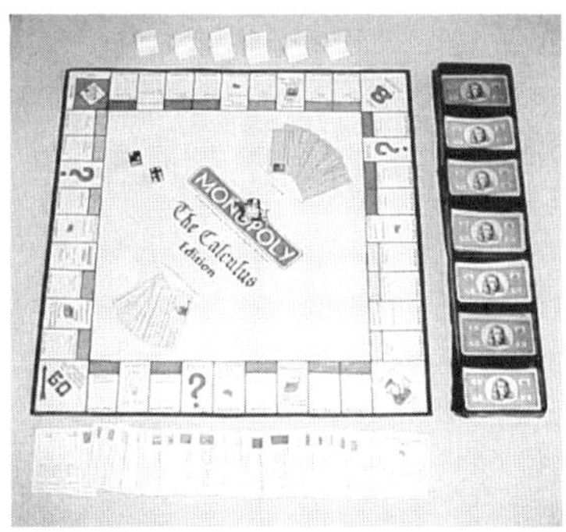

C8-2D Calculus project: Monopoly—The Calculus edition (colour plate xxxii)

Summary

Teachers are encouraged to make connections between the disciplines (i.e., for students), between the child and the educator (i.e., with students), and between the learners themselves (i.e., among students). All three of these connection types—teaching and learning between the lines—are regularly practised within a classroom that involves rich learning experiences (lessons/activities/extended projects). While the extended learning experiences, or projects, presented in the appendices focus on one particular method of integrating mathematics and visual art, be it understood that any discipline (and often more than just one) can likewise be creatively integrated in curriculum planning and implementation. In Chapter 1, we presented a brief overview of several curriculum-planning models, including the constructivist approach, problem-based learning, inquiry-based learning, the interdisciplinary model, brain-based learning, differentiated instruction, and universal design for learning. We contend that "whichever curriculum organizers are used—a form of one [already] described, or organizing structures yet unnamed—educators should constantly re-examine and evaluate their learning and teaching" (Naested, 2010, p. 13).

We, Irene and Dan, wish you many exciting and rewarding classroom interactions as your students are challenged by these and other rich learning experiences of your own (or their own) design. In the book, we have also attempted to demonstrate why such curriculum connections between mathematics and visual arts have been, and should continue to be, celebrated in contemporary classrooms and alternative educative contexts.

In the words of Albert Einstein, "It is the supreme art of the teacher to awaken joy in creative expression and knowledge." We believe that teaching and learning between the lines enables such two-fold awakenings for students and for teachers alike. We hope that you enjoy the journey.

References

Ascher, M. (1991). *Ethnomathematics: A multicultural view of mathematical ideas.* New York, NY: Chapman & Hall.

Attenborough, D., Pattison, E., Patsiatzis, T., & Muller, E. (1997). *Mathematics and visual art: An integrated programme for junior, intermediate and secondary levels inspired by the work of M. C. Escher.* St. Catharines, ON: Brock University.

Austin, S., Wadlington, E., & Bitner, J. (1992). Effect of beliefs about mathematics on math anxiety and math self-concept in elementary teachers. *Education, 112*(3), 390–396.

Ball, P. (1999). *The self-made tapestry: Pattern formation in nature.* Oxford: Oxford University Press.

Barth, P. (1995). Unforced fits between disciplines. *Education Digest, 61*(1), 34–36.

Bartuska, T., & Young, G. (Eds.). (1994). *The built environment: Creative inquiry into design and planning.* Menlo Park, CA: Crisp Publications.

Beane, J. (1993). Problems and possibilities for an integrative curriculum. *Middle School Journal, 25*(1), 18–23.

Bickley-Green, C. A. (1995). Math and art curriculum integration: A post-modern foundation. *Studies in Art Education. 37*(1), 6–18.

Binchy, M. (1992). *The copper beech.* London: Orion Books.

Blackwell, W. (1984). *Geometry in architecture.* New York, NY: John Wiley & Sons.

Blomfield, S. (2000). Math phobia can be beaten! *Ontario Mathematics Gazette, 38*(3), 5–7.

Bradley, J. (1971). *Introduction to Ruskin.* Boston, MA: Houghton Mifflin.

Brady, M. (1993). Single-discipline schooling. *Phi Delta Kappan, 74*(6), 438–443.

Broudy, H. S., & Palmer, J. R. (1965). *Exemplars of teaching method.* Chicago, IL: Rand McNally.

Burke, E. (1909). Proportion not the cause of beauty in the human species. In E. Burke, *A philosophical inquiry into the origin of our ideas of the sublime and beautiful.* Harvard Classics (Vol. 24, Part 2).

Campbell, S. (2010). *Growing patterns: Fibonacci numbers in nature.* Honesdale, PA: Boyds Mills Press.

Case, R. (1994). Our crude handling of educational reforms: The case of curricular integration. *Canadian Journal of Education, 19*(1), 80–93.

Clarke, J., & Agne, R. (1997). *Interdisciplinary high school teaching: Strategies for integrated learning.* Boston, MA: Allyn & Bacon.

Coelho, P. (1993). *The alchemist.* San Francisco: Harper Collins.

Coleman, L., & Huyghe, P. (1999). The field guide to Bigfoot, Yeti and other mystery primates worldwide. New York, NY: Avon Books.

Colman, S., & Coan, A. (1971). *Nature's harmonic unity: A treatise on its relationship to proportional form.* New York, NY: Benjamin Blom.

Cormack, W. (1991). On the struggle for school-based arts. *Design for Arts in Education, 92*(3), 40–45.

D'Ambrosio, U. (1985). Ethnomathematics and its place in the history and pedagogy of mathematics. *For the Learning of Mathematics, 5*(1), 44–48.
Davis, G. A., & Rimm, S. B. (1994). *Education of the gifted and talented* (3rd ed.). Needham Heights, MA: Allyn & Bacon.
De LaCroix, H., & Tansey, R. G. (1986). *Gardner's art through the ages* (8th ed.). Toronto, ON: Harcourt, Brace, & Jovanovich.
Delisle, R. (1997). *How to use problem-based learning in the classroom.* Alexandria, VA: Association for Supervision and Curriculum Development.
Dilke, O. (1987). *Mathematics and measurement.* London, UK: British Museum Publications.
Doczi, G. (1981). *The power of limits: Proportional harmonies in nature, art and architecture.* Boulder, CO: Shambhala Publications.
Donald Duck in mathmagic land. (1960). [Video]: Disney.
Donaldson, S. (2003, November). A driftwood mystery. *Pacific Yachting*, 66.
Dorn, C. M. (1994). *Thinking in art: A philosophical approach to art education.* Reston, VA: National Art Education Association.
Douady, S., & Couder, Y. (1992). Phyllotaxis as a physical self-organized growth process. *Physical Review Letters, 68*(13), 2098–2101.
Drost, J. (1999). The vortex tessellation. *The Mathematics Teacher, 92*(4), 286–289.
Dunn, R., & Dunn, K. (1978). *Teaching students through their individual learning styles: A practical approach.* Reston, VA: Reston Publishing Company.
Dunphy, P. (2004, January 18). Q & A, *Calgary Herald*, p. F8.
Edwards, B. (1999). *The new drawing on the right side of the brain.* New York, NY: Penguin Putnam.
Eisner, E. (1991). Structure and magic in discipline-based art education. In D. Thistlewood (Ed.), *Critical studies in art and design education* (pp. 14–26). Portsmouth, NH: Heinemann.
Eisner, E. (1999). Does experience in the arts boost academic achievement? *Clearing House, 72*(3), 143–49.
Elam, K. (2001). *Geometry of design: Studies in proportion and composition.* New York, NY: Princeton Architectural Press.
Feldman, E. (1967). *Art as image and idea.* Englewood Cliffs, NJ: Prentice Hall.
Fimmel, R. (n.d.). NASA Pioneer Project. Ames Research Center. Retrieved from http://history.nasa.gov/SP-349/ch9.htm
Fiore, G. (1999). Math-abused students: Are we prepared to teach them? *The Mathematics Teacher, 92*(5), 402–405.
Fisher, D. (2006, May 18). The art of weather, Group of Seven style. *Calgary Herald*, p. A19.
Flatland: The movie. (2007) [Video]: Flatworld Productions.
Fogarty, R. (1991). Ten ways to integrate curriculum. *Educational Leadership, 49*(2), 61–65.
Frank, F. (1993). *Zen seeing, Zen drawing: Mediation in action.* New York, NY: Bantam Books.
Freedman, K. (1987). *Art education as social production: Culture, society, and politics in the formation of curriculum.* Bristol, PA: Falmer Press.
Fry, P. (Ed.). (1994). *Alex Colville: Embarkation–The genesis of a painting.* Montreal, QC: The Montreal Museum of Fine Arts.

Fyfield, J., & Blane, D. (1995). *The Nelson Canadian school mathematics dictionary*. Toronto, ON: Nelson.
Gardner, H. (1983). *Frames of mind: The theory of multiple intelligences*. New York, NY: Basic Books.
Gardner, H. (2000). *The disciplined mind: Beyond facts and standardized tests, the K–12 education that every child deserves*. Toronto, ON: Penguin Books.
Gates, J. (1989). Consider the earth: Environmental activities for grades 4–8. Englewood, CO: Teacher Ideas Press.
Gerstner, K. (1986). *The forms of color: The interaction of visual elements*. Cambridge, MA: MIT Press.
Golen, F. (1999, December 31). Albert Einstein: Person of the century. *Time*, 24–27.
Grauer, K. (1991). Integrating art. *British Columbia Art Teachers' Association Journal for Art Teachers, 31*(1), 23–27.
Griffiths, P. A. (2000). Mathematics at the turn of the millennium. *The American Mathematical Monthly, 107*(1), 1–14.
Hall, E. T. (1966). *The hidden dimension*. New York, NY: Doubleday.
Hanley, B. (1994). Canadian arts education: A critical analysis of selected elementary curricula. *Canadian Journal of Education, 19*(3), 197–214.
Harim, E. & Love, J. (1998, May). A tree-mendous learning experience: Promote interdisciplinary inquiry with a tree. *Science and Children*, 26–31.
Hoffer, A. (1977). *Geometry and visualization: Mathematics Resource Project*. Palo Alto, OR: Creative Publication.
Hope, M. (2005, July 16). Zero Energy Plan for housing. *Calgary Herald*, p. J4.
Hubbard, G. (1986). *Art in action*. San Diego, CA: Coronado Publishers.
IBM Canada Ltd. (1983). *I, Leonardo: A journey of the mind*. Educational packet. Markham, ON: Author.
Irwin, R. L., & Reynolds, J. K. (1995). Integration as a strategy for teaching the arts as disciplines. *Arts Education Policy Review, 96*(4), 13–20.
Jacobs, H. (1970). *Mathematics: A human endeavor*. San Francisco, CA: W. H. Freeman and Company.
Jacobs, H. H. (Ed.). (1989). *Interdisciplinary curriculum: Design and implementation*. Alexandria, VA: Association for Supervision and Curriculum Development.
Janson, H. W. (1991). *History of art*. New York, NY: Abrams.
Jardine, D. (1990). To dwell with a boundless heart: On the integrated curriculum and the recovery of the earth. *Journal of Curriculum and Supervision, 5*(2), 107–120.
Jarvis, D. H. (2001). *Learning between the lines: A syncretistic experiment in mathematics and visual arts education*. Unpublished master's thesis, Nipissing University, North Bay, ON.
Jarvis, D. H. (2002). *De divina proportione:* The art and science of the golden ratio. *Canadian Art Teacher, 2*(1), 17–23.
Jarvis, D. H. (2007). Mathematics and visual arts: Exploring the golden ratio. *Mathematics Teaching in the Middle School, Focus Issue: Mathematics and the Arts, 12*(8), 467–473.
Jarvis, D. H. (2008). Thinking outside the rectangular prism: Fostering problem-based mathematics learning. *Ontario Mathematics Gazette, 47*(2), 23–28.

Jarvis, D. H. (2009). *Parametric creativity: Curriculum negotiation and professional development models in mathematics education.* Köln, Germany: LAP Lambert.
Jarvis, D. H., & Namukasa, I. (2009). Geometry, geography, and equity: Fostering global-critical perspectives in the mathematics classroom. *Delta-K, 46*(2), 18–26.
Jarvis, D. H. (2010). Portraiture as epistemological encounter: Knowing through skilful observation. In M. Blatherwick & H. Pearse (Eds.), *What works: Innovative strategies for teaching art* (pp. 103–106). Victoria, BC: Canadian Society for Education through Art.
Jarvis, D. H., & Franks, D. (2011). "Messy time" transition to reform-based mathematics teaching and learning. *Ontario Mathematics Gazette, 49*(4), 28–34.
Jensen, E. (1988). *Teaching with the brain in mind.* Alexandria, VA: Association for Supervision and Curriculum Development.
Johnson, D. (1974). *Excursions in outdoor measurement.* Portland, ME: J. Westin Walsh.
Johnson, D., & Glenn, W. (1961). *The world of measurement.* St. Louis, MO: Webster.
Kemp, M. (1996). Doing what comes naturally. *Art Journal, 55*(1), 27.
Kohl, M., & Gainer, C. (1996). *MathArts: Exploring math through art for 3–6 year olds.* Beltsville, MD: Gryphon House.
Kron, J. (1983). *Home-psych: The social psychology of home decoration.* New York, NY: Clarkson N. Potter.
Le Corbusier. (2008). *Towards an Architecture.* London, UK: Frances Lincoln. (original work published 1927).
Lesmoir-Gordon, N., Road, W., & Edney, R. (2000). *Introducing fractal geometry.* Cambridge, UK: Icon Books.
Life by the numbers. (2006). [Video Series]: Public Broadcasting Corporation.
Livio, M. (2002). *The golden ratio: The story of phi, the world's most astonishing number.* New York, NY: Broadway Books.
Mandelbrot, B. (1983). *The fractal geometry of nature.* New York, NY: W. H. Freeman.
Marieb, E. (1987). *Human anatomy and physiology: Laboratory manual* (2nd ed.). Meneo Park, CA: Benjamin-Cummings.
Marshall, S. (1970). *An experiment in education.* Cambridge, UK: Cambridge University Press.
Martel, Y. (2001). *Life of Pi.* Toronto, ON: Random House.
Martienssen, H. (1976). *The shape of structure.* London, UK: Oxford University Press.
Maslow, A. (1970). *Motivation and personality* (2nd ed.). New York, NY: Harper and Row.
Mason, T. C. (1996). Integrated curricula: Potential and problems. *Journal of Teacher Education, 47*(4), 263–271.
McCoy, L. P. (1992). Correlates of mathematics anxiety. *Focus on Learning Problems in Mathematics, 14*(2), 51–57.
Metcalfe, L. (n.d.). *Louvre pyramid: Reference.* Retrieved from http://www.thefullwiki.org/Louvre_Pyramid
Naested, I. (1991). Toward the forgotten thoughtfulness. In W. Shute & S. Gibb (Eds.), *Students of thought* (pp. 75–88). Calgary, AB: Detselig.
Naested, I. (1993, March/April). When I was a child. *The ATAW Magazine,* 6–8.
Naested, I. (1998). *Art in the classroom: An integrated approach to teaching art in Canadian elementary and middle schools.* Toronto, ON: Harcourt Brace Canada.

Naested, I. (2002, reprinted 2010). *Art in the classroom: An integrated approach to teaching art in Canadian elementary and middle schools*. Toronto, ON: Thomson Nelson.
Naested, I., Potvin, B., & Waldron, P. (2004). *Understanding the landscape of teaching*. Toronto, ON: Pearson.
Naisbitt, J., & Aburdene, P. (1990). *Megatrends 2000: Ten new directions for the 1990s*. New York, NY: William Morrow & Co.
National Council of Teachersof Mathematics. (2000). *Principles and standards for school mathematics*. Reston, VA: Author.
Newman, R., & Boles, M. (1992). *Universal patterns: The golden relationship of art, math, and nature*. Bradford, MA: Pythagorean Press.
Ontario Ministry of Education. (2005). *The Ontario curriculum, grades 1–8: Mathematics, revised*. Toronto, ON: Queen's Printer for Ontario.
Pedoe, D. (1976). *Geometry and the liberal arts*. Markham, ON: Penguin Books.
Phi-Nest. (n.d.). Human beauty: Beauty is in the phi of the beholder. Retrieved from http://goldennumber.net/beauty.htm
Plato. (1987). *The Republic* (2nd ed.). Translated (1955) by S. Lee. London, UK: Penguin Books.
Posamentier, A., & Lehmann, I. (2007). *The fabulous Fibonacci numbers*. Amherst, NY: Prometheus Books.
Posamentier, A. S., & Stepelman, J. (1990, November). Teaching secondary school mathematics: Techniques and enrichment units. *Mathematics Teacher*, 684–685.
Preble, D. (1973). *Man creates art creates man*. Berkeley, CA: McCutchan.
Ritter, D. (1995). *Math art: Featuring 33 innovative art projects*. Cypress, CA: Creative Teaching Press.
Ross, C. (1996). *Squares, shapes in math, science and nature*. Toronto, ON: Kids Can Press.
Rowland, A. (1990). *Bauhaus source book: Bauhaus style and its worldwide influence*. Oxford, UK: Phaidon.
Runion, G. E. (1990). *The golden section*. Palo Alto, CA: Dale Seymour Publications.
Schattschneider, D. (1990a). *M. C. Escher: Visions of symmetry*. New York, NY: Freeman & Co.
Schattschneider, D. (1990b). *Visions of symmetry, notebooks, periodic drawings and related works of M. C. Escher*. New York, NY: W. H. Freeman & Company.
Schattschneider, D. (1993). The fascination of tiling. In M. Emmer, *The visual mind: Art and mathematics* (pp. 146–157). Cambridge, MA: MIT Press.
Schneider, M. (1994). *The beginner's guide to constructing the universe: The mathematical archetypes of nature, art and science, Voyage from 1 to 10*. New York, NY: Harper Collins.
Schramm, S. (1997). *Related webs of meaning between the disciplines: Perceptions of secondary students who experienced an integrated curriculum*. (ERIC Document Reproduction Service No. ED 407258)
Scott, P. (1974). *Discovering curves and surfaces*. Melbourne, Australia: Cheshire Publishing.
Selke, D. H. (1999). Geometric flips via the arts. *Teaching Children Mathematics, 5*(6), 379–383.
Seymour, D., & Britton, J. (1989). *Introduction to tessellations*. Palo Alto, CA: Dale Seymour Publications.
Shields, C. (2003). *Unless, a novel*. Toronto, ON: Vintage Canada.

Shlain, L. (1991). *Art and physics: Parallel visions in space, time and light.* New York, NY: Morrow.
Sitomer, H., & Sitomer, M. (1974). *Spirals.* New York, NY: Thomas Y. Crowell.
Small, M. (2008). *Making math meaningful to Canadian students, K–8.* Toronto, ON: Nelson.
Stewart, I. (1995). *Nature's numbers: The unreal reality of mathematical imagination.* New York, NY: Basic Books.
Stewart, I., & Golubitsky, M. (1992). *Fearful symmetry: Is God a geometer?* Oxford, UK: Blackwell.
Stocker, D. (2007, March). Putting "pizza party" math to rest. *Teacher Newsmagazine, 19*(5). Retrieved from http://bctf.ca/publications/NewsmagArticle.aspx?id=11512
Sylwester, R. (1995). *A celebration of neurons: An educator's guide to the human brain.* Alexandria, VI: Association for Supervision and Curriculum Development.
Taylor, A. (1974). *Math in art.* Hayward, CA: Activity Resources Company.
Taylor, B. D. (1974). *Design lessons from nature: Drawing and design lessons based on the scientific study of natural forms.* New York, NY: Watson-Guptill.
Thames, M. H., & Ball, D. L. (2010). What math knowledge does teaching require? *Teaching Children Mathematics, 17*(4), 220–229.
The story of 1. (2005). [Video]: British Broadcasting Corporation.
The story of math. (2010). [Video]: Athena.
Tomlinson, C. A. (1999). *The differentiated classroom: Responding to the needs of all learners.* Alexandria, VA: Association for Supervision and Curriculum Development.
Trapp, D. (n.d.). Crypto-bipedal-Primatotolgy: Sasquatch. Retrieved from http://www.xmarks.com/site/ourworld.compuserve.com/homepages/dtrapp/bigfoot.htm
Verstockt, J. (1987). *The genesis of form: From chaos to geometry.* London, UK: Muller, Blond & White.
Vitz, P. C., & Glimcher, A. B. (1984). *Modern art and modern science: The parallel analysis of vision.* New York, NY: Praeger.
Weyl, H. (1969). *Symmetry.* Princeton, NJ: Princeton University Press.
Whitford, F. (1984). *Bauhaus.* London, UK: Thames and Hudson.
Wiggins, G., & McTighe, J. (1998). *Understanding by design.* Alexandria, VA: Association for Supervision and Curriculum Development.
Williams, E. (1988). *Survival mathematics: Basic math to help you cope.* Woodbury, New York, NY: Barron Educational Series.
Woodbury, S. (1998). Rhetoric, reality, and possibilities: Interdisciplinary teaching and secondary mathematics. *School Science and Mathematics, 98*(6), 303–312.

Appendix A

Glossary of Key Mathematics Terms

algorithm: A step-by-step procedure for doing calculations
altitude: The height at the uppermost point of a triangle, or other shape
angle: Amount by which something turns or rotates and is measured in degrees
approximation: A guess which might involve arithmetic, measuring or visual perceptual skills
arc: Curved line of a circle
arch: A curved, symmetrical structure spanning an opening and serving as a support
architecture: The art of designing and constructing buildings and other environmental features
area: A measure of the total surface of an object
arithmetic: The adding, subtracting, multiplying, and dividing of numbers
asymmetry: A balance made up of unlike shapes or other design elements on each side of an imaginary central axis (e.g., human lungs, fiddler crab)
axis: Line through the middle of a shape, real or imagined
balance: A weight or force counteracting the effect of another; a weighing device
base: The bottom edge on which a shape rests
bilateral symmetry: A mirror image, or having reflective symmetry
box: Having a flat bottom and four sides
calendar: A system for counting the years, months, and days
circle: There are 360° in a circle, 180° in half a turn, and 90° in a quarter circle (or right angle)
circumference: The edge of a curved shape such as a circle; the distance around a circle
clock: A device to measure time featuring a 12-hour cycle (a.m., ante meridiem, and p.m., post meridiem), 60 minutes per hour, and 60 seconds per minute; are analog (circular face) and digital varieties
clockwise: Moving around in the same direction as the hands on a clock, as opposed to counter-clockwise which is moving in the opposite direction
columns: Numbers, lines, or shapes that are vertical (up-and-down); as opposed to rows, which are horizontal
comparison: Observing objects to establish similarities or dissimilarities
compass: An instrument consisting of two legs joined at the top by a pivot; used for drawing circles and arcs, and for transferring measurements
concave: The inward curve of a smooth surface (e.g., the inside of a cereal bowl)
concentric circles: Circles lying in the same plane and having a common centre
cone: A circle-based prism; a three-dimensional figure with a pointed top and circular base
congruent: Two figures that are exactly the same size and shape
conic sections: Four types of curves formed by the intersection of a plane and a double right circular-based prism or cone (namely, the circle, ellipse, parabola, and hyperbola)
convex: The outward curve of a smooth surface (e.g., the outside of a ball)

coordinates: Two numbers that denote a specific position of a point related to a set of axes

coordinate geometry: Two lines drawn perpendicular to each other (x- and y-axis); the position of any point can be given by its distance from each of these lines; the Cartesian plane, named after René Descartes, is among the most popular coordinate systems used in mathematics and features an x-axis, y-axis, four quadrants, and a centre (the origin)

cross-section: A slice through a solid shape

cubes: Boxlike three-dimensional figures that have six equal, square faces

cylinder: A three-dimensional figure with two congruent, parallel, circular faces and one curved surface (e.g., soup can, water barrel)

degree: A unit on a measuring scale; also used to measure angles of a circle (360°)

diagonal: Extending from one corner to an opposite corner in a four- or greater-sided figure

diagram: A drawing, sketch, plan, or chart that makes something easier to understand

diameter: A line that cuts a circle in half

dilatation (or dilation): A transformation that enlarges or reduces a shape by a scale factor to form a similar shape

dimensions: The length of an object in different directions (e.g., height, width, depth)

dodecahedron: One of the five Platonic solids having twelve regular pentagon plane faces; in ancient Greek cosmology, associated with "divine" due to its golden proportions

edge: A straight or curved line that marks the boundary between two faces in three-dimensional geometry

ellipse: A plane curve that results from the intersection of a cone by a plane in a way that produces a closed curve (one of the four conic sections); note that a circle is a special case ellipse, where the two focal points are coincident (same)

equiangular: Having all angles identical in degrees

equilateral: Having all sides identical in length (e.g., an equilateral triangle)

estimate: A rough decision or judgment on a number, amount, value, or size of an object or answer; often referred to as an informed conjecture

ethnomathematics: A branch of mathematics education dealing with the study of the relationship between mathematics and culture

face: One of the flat sides in a solid three-dimensional figure

Fibonacci sequence: A sequence of numbers in which each following number is the sum of the two preceding numbers; 0, 1, 1, 2, 3, 5, 8, 13, 21, 34, 55, . . . ; Fibonacci spirals occur in nature in certain plants wherein two spirals, one winding clockwise and the other counterclockwise, fit together and follow the Fibonacci number sequence

flip (reflection): To turn over or to reflect a mathematical object about a given point or axis; one of four basic actions in transformational geometry

fractal: Benoit Mandelbrot (1924, Poland) coined the term *fractal* for geometric figures that are highly irregular or fragmented spatial patterns often found in nature. The word comes from the Latin word *fractus*, meaning "to break"

fraction: A part of a whole number (e.g., ¼, where the numerator [top] denotes one of the four possible parts noted in the denominator [bottom]); fractions can be interpreted as part of a whole, part of a set, as division among groups, or as a ratio

frequency: The number of times an event or outcome occurs; recorded in frequency tables

geodesic: The shortest path connecting two points on a surface; a geodesic dome is a spherical or partial-spherical shell structure based on a network of great circles (geodesics) on the surface of a sphere

geometric shapes: In geometry, two-dimensional (flat) objects such as an ellipse, triangle, rectangle, pentagon, trapezoid, etc. (Note: squares are special case rectangles and circles are special case ellipses.)

geometric figures: In geometry, three-dimensional objects such as a sphere, prism, or pyramid (Note: a cylinder is actually a circle-based prism; the cone a circle-based pyramid.)

geometry: The branch of mathematics that deals with lines, curves, angles, and shapes; geometrical shapes are made from combinations of straight lines, circles, and arcs

glide reflection: A transformation consisting of a translation or glide along a line as well as a flip or reflection in the slide axis (e.g., footprints)

golden ratio (section; mean): In mathematics and the arts, two unequal line lengths are in the golden ratio if the ratio of the sum (total) of the lengths to the larger length is equal to the ratio of the larger length to the smaller length; the golden ratio is an irrational mathematical constant, approximately 1.6180339887

golden (divine) proportion: In mathematics, two equal ratios are known as a proportion. The golden proportion occurs when the ratio of the smaller section of a divided line compared to the larger section, is equal to the ratio of the larger section compared to the entire line length ("short is to long, as long is to whole"). When actual measured lengths are inserted into the proportion, the quadratic formula produces (see p. 61 for equation) or the irrational number 1.6180339887, commonly referred to as "phi" (φ).

graph: A diagram that shows a collection of data, or how one set of data changes with regards to another set of data (e.g., average monthly rainfall; height vs. arm span)

grid: An organized system with evenly divided and equally spaced squares on a figure or flat surface, often using a set of numbers and/or letters to designate position

half: Cut or divided into two equal parts

height: The highest part or point; the distance from the bottom to the top of something standing upright

helix: A three-dimensional version of the Archimedean spiral, such as a candy cane strip, corkscrew, or coil springs; a curve that lies on a cylinder or cone in a spiral-like manner

hemisphere: Half of a sphere (e.g., the western hemisphere includes North America)

heptagon: A polygon featuring seven angles and seven sides

hexagon: A polygon featuring six angles and six sides

horizon: Where the land and sky appear to meet; horizontal, therefore, means a line or object lying in the same direction as the observed horizon

icosahedrons: A regular (made up of equilateral triangles) polyhedron featuring twenty faces; in ancient Greek cosmology, associated with the basic element of water

infinity: Counting on forever; usually represented by the sign ∞

intersect: To share common points

intersecting lines: Two or more lines that cross each other

latitude: Imaginary circles drawn around the earth between the equator and the north and south poles

length: The distance between the two points

line: A geometrical object formed by a straight set of points that extends to infinity in both directions, having length but no depth or width; a set of the points whose coordinates satisfy a given linear equation on the two-dimensional Cartesian plane or in three-dimensional Euclidean space.

linear equation: An algebraic representation of a linear relationship that appears as a straight line when graphed on a coordinate system

line length: The shortest distance between two points on a flat plane

line of symmetry: The line that divides a shape into two equal parts, or that produces a shape and its mirror image

longitude: Imaginary lines drawn around the earth, each passing through both the North and South poles

mandala: A radial design often symbolic of the seasons, life and death, calendar of events, earth or the universe; traditionally used in Hindu or Buddhist meditative images

map: A drawing or diagram of a place drawn to scale

mass: The amount of matter in an object, usually measured in grams or kilograms

mean: One measure of central tendency, often referred to as the "average"; found by dividing the sum of a set of numbers by the number of numbers in the set

metric: An international decimalised (using powers of ten) system of measurement, first adopted in France (1799) and now the basic system of measurement used in almost every country (except the United States); length is measured in centimetres, metres, kilometres, etc.; mass in grams, kilograms, etc.; volume in litres, millilitres, etc.

mobile: A construction made with parts, generally suspended, that can move through use of air currents or electricity

model: A miniature representation of a building or object, or a concrete/visual representation of an abstract idea/situation; models in mathematics often now involve some form of hands-on manipulative or software representation

modular: Constructed with a standard unit or dimension

money: Coins, notes, or other tokens used as a shared "currency" for exchanging goods or services

motion: An act or process of changing position

negative space: The space around a positive shape

net: A pattern that can be folded to make a three-dimensional figure

numbers: Symbols used to denote quantity (e.g., Hindu-Arabic [1, 2, 3, . . .] and Roman [I, II, III, . . .] numerals)

octagonal: A polygon with eight sides

octahedron: A polyhedron with eight faces; the "regular" octahedron is one of the five Platonic Solids

origin: The point at which the coordinate axes intersect (0,0)

oval: An egg-shaped object (not to be confused with the conic section ellipse having perfect symmetry)

parabola: A curved line, or conic section, described by the quadratic function ($y = x^2$)

parallel lines: Lines in the same plane that do not intersect

pattern: A pattern is a special arrangement of numbers or shapes

pentagon: A polygon having five angles and five sides; a regular pentagon has five equal sides
pentagram: A five-pointed figure or star formed from within a regular pentagon by joining the vertices
percent: One part of a whole that has been divided into 100 parts; a percent compares a number to 100 (e.g., 77% means "77 out of 100" or 77/100)
perimeter: The edge, boundary, or rim of an area or figure; the length of the boundary
perpendicular: Two lines that meet or cross at a right angle of 90°
phi (symbol φ): A special irrational number, (see p. 61 for equation) named in honour of Phidias, a Greek sculptor; also referred to as the golden mean, section, ratio, or "divine proportion"
phyllotaxis: Leaf arrangement; the study of geometric and numerical patterns in plants
pi (symbol ϖ): The ratio of the circumference to the diameter of any circle (equal to 3.141592 . . .)
plan: A drawing or design for an object or building
plane: Any flat, two-dimensional surface; plane shapes have length and width but no thickness
Platonic solids: Plato (427–347 BC), a Greek philosopher, founded a school in which members studied mathematical phenomena such as the five regular polyhedra, which became know as the five Platonic solids (tetrahedron, hexahedron (cube), octahedron, icosahedron, and dodecahedron); in Greek cosmology, these solids were considered to be related to the four basic elements of nature (earth, fire, air, water), with the dodecahedron—based on pentagons and involving many iterations of the golden section—considered to be related to the "divine" or metaphysical realm
point: A mark on a graph, grid, or map indicating location or position
polygon: A closed, flat shape formed by three or more line segments (e.g., triangle, quadrilateral, pentagon, octagon, chiliagon, myriagon, megagon)
polyhedron: A three-dimensional figure that has polygons as its faces
position: The order, place, or location of an object
prism: A three-dimensional figure with two parallel and congruent bases; a prism is named by the shape of its bases (e.g., triangular prism); a glass triangular prism can separate white light into the colours of the rainbow (red, orange, yellow, green, blue, indigo, violet)
projection: The process or technique of reproducing a spatial object upon a plane or curved surface by projecting its points; a view of a building or architectural element
proportion: An equation showing equivalent ratios in fraction form
protractor: An instrument used for measuring and marking out angles with a scale that is marked in degrees
pyramid: A polyhedron whose base is a polygon and whose other faces are triangles that meet at a common vertex; a massive stone structure built in ancient Egypt as a tomb for a deceased Pharaoh
quadrant: One-quarter of the Cartesian plane, bounded by the coordinate axes; note that Quadrant I begins in positive x and y values (top-right), Quadrant II in negative x and positive y values (top-left), and so on in a counter-clockwise manner for the four quadrants
quadrilateral: A polygon shape with four sides
radius: A line segment with endpoints at the centre of a circle and at a point on its circumference
radial: Arranged or having parts like rays originating from a centre (e.g., mandala)
random: Made or chosen by chance

rate: A comparison of two measurements with different units, such as distance and time (e.g., km/hr)

ratio: The relationship in quantity, amount, or size between two or more things; note that two equivalent ratios (often written as fractions) are known as a proportion

rectangle: A quadrilateral in which opposite sides are equal and all interior angles are right angles

reflection (flip): Transformation of a geometric shape across an axis that results in a congruent, mirror image of the original shape

regular polygon: A closed figure in which all sides are of equal length and all interior angles are equal; Greek prefixes used to denote number of sides: (e.g., tri (3), tetra (4), penta (5), hexa (6), hepta (7), octa (8), ennea/nona (9), and deca (10))

rhombus: A parallelogram (four-sided shape) with equal sides

Roman numerals: A number symbol system that was replaced in the western world by the Hindu-Arabic number system owing to the latter's ability to more easily handle advanced calculations; basic Roman numerals are I (1), II (2), III (3), IV (4), V (5), VI (6), VII (7), VIII (8) IX (9), X (10), XI (11), XX (20), XL (40), L (50), C (100), M (1000)

rotation (turn): A transformation that turns a shape about a fixed point to form a congruent shape

rotational symmetry: An object that looks the same after a certain amount of rotation (less than 360°)

row: Numbers, lines, or shapes that are arranged horizontally as opposed to columns. which are arranged vertically (up-and-down)

ruler: A straight-edged stick with a marked scale used to measure lengths and draw straight lines

scale: A sequence of numbers associated with marks that subdivide an axis on a graph; also, an instrument used in measuring an object's mass

scale drawing: A drawing in which the lengths are proportionally enlarged or reduced from the original lengths according to a given ratio (e.g., the scale architectural drawings for a new building)

scatter plot: A graph of plotted points that show the relationship between two sets of data

skeleton: A model that shows only the edges and vertices of a three-dimensional figure, i.e., without the faces

sphere: A perfectly round ball or globe in which all surface points are equally distant (one radius) from a centre point

spiral: A curved line that moves away from, but also around, a fixed point at a changing distance

square: A special case rectangle in which all four sides are equal

sundial: An instrument used to determine the time of day by the shadow of a gnomon cast on a horizontal plate or cylindrical surface

symmetry: The geometric property of being balanced about a point, a line, or a plane

tangent: A straight line that touches a curve but does not cross it

tessellation: A tiling pattern in which shapes are repeated to completely cover a two-dimensional surface without overlap or spaces left between repeated shapes (e.g., wallpaper patterns, paving stones)

tetrahedron: A polyhedron with four faces; a regular tetrahedron has four equilateral triangles as faces, and is one of the five Platonic solids

three-dimensional: A solid figure that has length, width, and height

transformation: A mapping of one geometrical figure to another figure according to a specific rule or set of rules (e.g., there are four main transformations: translations/slides, rotations/turns, reflections/flips, and dilatations/enlargements or reductions)

translation (slide): A transformation that moves every point on a shape the same distance, in the same direction, to form an identical shape at a new position

triangle: A polygon with three sides

two-dimensional: A shape that has length and width but no height (thickness)

vertex: The common endpoint of two or more line segments; in three dimensions, a corner

volume: The amount of space occupied by an object, measured in cubic units (e.g., cubic centimetres); capacity is often used synonymously with volume, but usually denotes an amount of liquid fill

Appendix B

Glossary of Key Visual Arts Terms

aesthetics: A branch of philosophy dealing with the nature, creation, and appreciation of beauty; the quality of attractiveness, or what is individually or socially considered visually pleasing

appliqué: Cut fabric pieces arranged, glued, or stitched to a larger piece of cloth

architecture: The art and science of planning and building structures

armature: In sculpture, the framework around which the sculpture is built

assemblage: A work of art that has been assembled instead of drawn, painted, modelled, or carved

asymmetry: A balance made up of unlike shapes on each side of the picture plane or a central axis

awareness: The first stage in any creative process, the ability to elicit in vivid detail the thoughts, perceptions, and feelings that comprise an experience

balance: A situation in which all parts of an artwork are arranged, either formally and informally, to provide visual symmetry to the work; formal balance occurs when equal or very similar elements are placed on opposite sides of an axis; informal balance occurs when the elements on either side of the axis are, in an overall sense, visually equal in size (e.g., one large object on one side, two medium-sized objects on the other side)

batik: Javanese meaning "wax writing," a wax-resist process of dying and decorating fabric

bilateral: Having or affecting two sides; two-sided

blind-contour drawing: Drawing while keeping one's eye on the object or form while the pencil delineates the form on paper (i.e., drawing without looking at one's drawing tool)

calligraphy: Elegant handwriting using the techniques developed by the Chinese centuries ago through simplification and/or elaboration of images

caricature (cartooning, characterization): Drawing objects, people, or animals by exaggerating their features or, in the case of animals, giving them human qualities

carving: A sculptural process that requires altering the appearance of a solid mass (e.g., wood, stone) by subtracting or cutting away (e.g., wood carving, marble sculpting)

casting: Reproducing a shape by filling a mould with material such as clay or plaster

centre of interest: A focal point in an artwork designed to attract and hold the viewer's attention
clay: A naturally-occurring aluminium silicate composed primarily of fine-grained minerals; used for sculpting three-dimensional works of art that are subsequently kiln-fired for permanence
collage: A technique of making art by pasting various images or materials on a base form; from the French verb *coller* meaning to paste or glue
collograph: A relief print made from a surface that has been built up by gluing or attaching shapes or objects
colour: An element of art/design that refers to the range of different wavelengths of light that can be seen by the human eye when an object reflects or emits light; four characteristics of colour are hue, value, intensity, and temperature
colour schemes or harmony: Specific colour selections placed together in a composition to produce a desired visual effect (e.g., complementary colours are considered harmonious)
composition: The organizing of forms and their qualities to create visual or auditory meaning or impressions within works of art (e.g., a song or instrumental ensemble, a dance or movement sequence, a play, or a visual composition)
concentric: A design having a common centre
continuous line drawing: Sketching the contour or outline of a subject, or parts thereof, without lifting the pencil or brush
construction: The assembling of materials in a building process
construction lines: Lightly drawn lines used to guide the construction of other key lines/objects in perspective drawing
contour drawing: Drawing the external edge or boundary of an object, resulting in a line drawing that is developed detail-by-detail to arrive at a completed form; emphasizes the co-ordination of the sense of touch with the sense of sight (hand/eye); coordinating what is seen with what is drawn
contrast/variety: Using the elements of composition to achieve opposition and interest; ways of adding interest to a composition by placing emphasis on a particular area and thereby avoiding excessive unity, harmony, or repetition
converging lines: Lines that meet at a vanishing point
cool colours: Colours such as blue, green, and violet, which produce the visual effect of objects receding
creativity: The ability to produce, initiate, or invent something on one's own, rather than to imitate or obey
cross-contour drawing: Taking contour drawing further to give roundness, or three-dimensionality, to objects (e.g., when drawing a human hand, artists draw the outside contour first, then the rounded shapes of fingers, etc., trying to "feel" their way across the various internal object contours)
cross-hatching: A drawing technique for shading using numerous crossed sets of parallel lines that are used in layers to create the effect of depth and three-dimensionality
curator: A person who collects and arranges works of art for an art show in a gallery or museum
decorated: Embellished or made more attractive
detail: Creates the illusion of depth in an artwork; uses the fact that greater detail can be seen on objects that are closer to the viewer, and less detail on objects at a distance

dimensions: The width, length, and height measurements of an object or shape; the fourth dimension is defined in physics as that of time, often depicted in artworks using symbols or combined chronological imagery

diorama: A constructed or designed miniature environment, often made as a scale replica of the original

distortion: Changing an object by deforming it, possibly to fit a defining shape or frame (e.g., Mondrian's trees filling an oval or circle)

distortion grid: A network of horizontal and vertical lines that are not parallel or evenly spaced. They can be regularly or irregularly stretched, patterned, or curved

dry brush: A painting technique in which strokes are made using a brush that is almost dry

elaboration: Decorating an artwork with dots, circles, paisley, etc. (e.g., Art Nouveau artworks)

elements of art/design: Fundamental components of artworks (e.g., line, shape, texture, form, space, value, and colour)

embossed printing: A process of pressing paper over a raised area or plate

emphasis: Focusing attention on one area of an artwork (e.g., using colours, unique lines/forms), thereby making it dominant in terms of the viewer's experience, and controlling the viewing sequence

etching or intaglio: Printmaking technique in which grooves are made on a plate surface, ink is forced into the grooves, the excess ink on the surface is wiped away, and then the image is transferred to prepared (i.e., dampened) paper by applying pressure to the plate, usually by means of a press

exaggerate: Drawing, painting, or sculpting part of an object in detail and then enlarging it, changing proportions, shapes, colours, sizes, etc. for a desired visual effect

figure-ground perception: Figures in a composition represent positive shapes or space, and the ground or area around the shapes represent negative space

frieze: A decorative band along the top of a wall, doorway, or archway

geometric drawing: A drawing using only geometric shapes

gesture drawing: A drawing done quickly to capture the essence of the subject's pose, rather than to study the finer details of anatomy; artists often use gesture drawing as a means of mental and physical warm up prior to a long drawing session

ghost drawing: A preparatory design and warm-up technique in drawing in which the artist quickly traces out the contours of the objects being drawn without actually touching the drawing tool to the paper

gouache: Opaque, water-based paint

grid: A network of horizontal and vertical lines; the use of a grid pattern may assist artists in accurately depicting a subject's proportions; the grid can be used for direct copying, enlargement, reduction, or distortion of the object/subject

harmony: A principle of design that describes how elements are combined to highlight their similarities and thereby produce a harmonious, visually-pleasing composition

horizon line: The line at which the sky and earth appear to meet, usually at the viewer's eye level

illusion of depth: A number of techniques have been developed over centuries to create the appearance of three-dimensional depth on a flat, two-dimensional artwork, a list of which would include the following: the use of foreground, middle ground, and background;

overlapping shapes; varying size; amount of detail; position on the picture plane; greying of colour; and linear perspective

illustration: A two- or three-dimensional image intended to explain or decorate (e.g., images found in books)

imagination: The ability to recall and project thoughts, feelings, and perceptions in a personal manner

intensity (of colour): Refers to the brightness of a colour; a colour is at full intensity only when it is pure and unmixed

interpretation: A personal response to the perceived meaning of a visual artwork

jewellery: Objects that are found or handmade to adorn parts of the human body or hair

kaleidoscope: Mirrored instrument containing bits of coloured glass reflecting a changeable pattern

line: An element of design, a line is the visual path (straight or curved) left by a point moving through space; lines are found in both natural and constructed objects; lines define space or are used to create decoration, pattern, or texture

lines of perspective: Lines that converge to one or more vanishing points; used by artists to create the illusion of depth; developed as a science during the Renaissance period

logo: An identifying symbol used to advertise a company or organization

macramé: Decorative knot tying

magnification: A technique in which a small detail of an object is enlarged, often used for abstraction in art

mandala: A graphic, mystic symbol of the universe (e.g., Hindu/Buddhist mandalas) that is typically in the form of a circle and that often involves intricate patterns and colour schemes

mask: A work of art generally made to be worn on the head, covering or partially covering the face to disguise or transform the wearer

mass: The amount of matter in an object, usually measured in grams or kilograms; the illusion of mass can be created on a two-dimensional surface by considering light, shade, and shadow (e.g., basic forms such as the sphere, prisms, and pyramids are defined by shading, which intimates mass)

mass drawing: Rendering the solidity of the subject using areas of tone or colour, without emphasizing lines or edges

medal: Objects generally made of metal and used to commemorate a person or event, or awarded for excellence or achievement

media: Materials used by artists to create works of art

metamorphosis: Drawing, painting, or sculpting an image in a progressive state of change, one frame at a time (e.g., a tree becomes a ballerina, a Swiss army knife becomes an automobile)

mirror image: A drawing, painting, or sculpture that is a reflection of the original; self-portrait drawing/painting often involves the use of double mirrors in order to reverse the image to match the viewer's perspective of the artist

mobile: A three-dimensional sculpture suspended in the air

modelling: A sculptural process that requires the rearranging of material through manipulation, sometimes adding or carving away (e.g., Plasticine, clay, playdough)

monoprint: An image that is drawn with inks or paint on a flat surface, and then only one print produced
mosaic tile: An artwork made with small pieces of inlaid coloured glass, clay tiles, or stone to form compositions
motif: A dominant, often stylized or simplified symbol or design; a theme that may be repeated
movement: A principle of composition using the elements to create viewing paths or visual direction, often leading the viewer to a planned focal point
mural: A large, flat two-dimensional composition (e.g., Michelangelo's Sistine Chapel ceiling paintings)
negative spaces: Shapes or spaces surrounding or inside an object that is considered the positive shape
optical illusions: Usually a non-objective artwork with an illusionary effect
organic: Biomorphic, free-form, or irregular shapes found in nature, often composed of curving exterior edges rather than straight lines
overlapping: A technique used to create the illusion of depth, where one shape is partly hidden from view by another shape (e.g., a tree behind another tree appears farther back)
papier-mâché: From the French term meaning "paper pulp," a technique used to create relief artworks and sculptures, where paper is torn in strips and soaked in a mixture of paste, glue, or flour and water and then applied to a three-dimensional wire object/model or flat surface
pattern: A visual repetition made from one or more elements of composition; elements are repeated in a regular or a irregular fashion
perception: The process by which one obtains information about the world through one's senses
picture plane: The two-dimensional surface on which a composition is created
point of view: Looking at an object from different perspectives (e.g., an ant's eye view, bird's eye view, front, back, top, sides, and from various angles)
portfolio: An organized collection of work in different media, that can take the form of a "process portfolio" with works in progress, or a finished product portfolio
portrait: A two- or three-dimensional artwork which attempts to create an exact likeness, or a modified/abstracted version, of a person or an animal
positive shape: Shapes or forms on a two-dimensional surface; the opposite of negative shapes/spaces are the areas surrounding the positive shape/space
principles of art/design: Organizing principles or guidelines regarding how the Elements of Design can be manipulated in artworks to achieve desired effects (e.g., balance, rhythm, emphasis, movement, and variety)
print: An image created and reproduced by hand, on paper or other surfaces using a printmaking technique (e.g., linocut, etching, silk screen); several prints are often produced by using the same image: this is referred to as a series, set, or edition
proportion: A principle of art/design that is concerned with the relationships between objects on the basis of size and number, or between parts of single object; in math, a proportion is two equal ratios
quilting: A craft in which layers of fabric and filler are stitched through with yarn or thread, and which often involves elaborate and colourful patterning
radial symmetry: A type of balance that may have four or more matching units that seem to

radiate or come out from the middle; radial balance occurs often in nature (e.g., flowers, orange segments)

realism: An art style in which the artist attempts to represent a subject as it appears naturally or realistically; photography/video has changed/diminished the relatively high value once placed on realism

relief: Works of art (paintings or sculptures) having areas that are raised above the surface; often only meant to be viewed from the sculpted side

repetition: A principle of art/design in which similar elements or images are used more than once in a work of art; repetition often produces the effect of movement or energy in a drawing/painting

reproduction: A copy of an original work of art; in many art schools, students reproduce, or copy, the works of masters in order to increase their appreciation of the techniques used, and to improve their skill

rhythm and movement: A principle of art/design in which the use of recurring elements causes the viewer to direct his/her view in a certain way, and which often add a sense of energy to the artwork

rubbings (frottages): An image made by rubbing drawing material on paper/fabric stretched over a textured surface

sculpture: Artworks that are three-dimensional, i.e., possessing height, width, and depth

shade: A darker version of a colour (low-value) made by adding black to that colour

shading: The use of light and shadow to create a sense of depth on a two-dimensional surface

shape: An element of art/design that describes the external outline, or form, of a two-dimensional image, and which can be geometric or organic, positive or negative

silhouette: The representation of an object filled in with solid colour, most often in black

simplification: Simplifying an image into more basic, geometric shapes; visually dividing an object into black, white, and grey areas (this can be done by squinting eyes, or blurring a projection to see tones)

skeleton: A support structure designed as the base of a clay or found-objects sculpture

sketchbook: A journal or learning log that collects class notes, preparatory drawings, ideas worth investigating, pictures, photocopies, poems, thoughts and research on art/artists

space: An element of art/design that refers to the area around, within, or between images (three-dimensional space created visually through techniques such as overlap, placement on plane, and amount of detail)

stamp print: A print made by stamping a found or handmade object in ink or paint, and then stamping the prepared object on paper or other surfaces

stencil: In serigraphy (silkscreen printing) an image is cut into a protective sheet of paper used as a block-out or mask and ink is applied through the mask to the surface underneath (e.g., paper, fabric)

stereotypical images: A symbol of minimal detail that is socially learned and understood (e.g., a sun made with a circle with radiating lines, a house made with a triangle on top of a rectangle or square)

still-life drawing: A drawing of a collection of non-living, immobile objects

stitchery: A craft in which needle and thread are used to decorate fabrics (also needlework, embroidery)

stylized: To represent an object or person according to a particular pre-conceived pattern or style
surrealist: A form of art in which a drawing, painting, or sculpture involves one or more common objects placed into uncommon environments or that are depicted with unusual proportions or qualities
symbols: Images or objects used in artworks to represent an organization, person, place, event, or idea
texture: The feel or appearance of a surface or substance
three-dimensional: A solid shape or figure that possesses length, width, and height (or depth); sculptures are three-dimensional in that they fill a three-dimensional space and can be seen in the round
tint: A derivative (high-value) of a colour made by adding white pigment
tone: The property of a colour that describes its degree of lightness or darkness; adding black pigment darkens or lowers the tonal value of a colour, whereas adding white pigment lightens or raises the tonal value of a colour
transformation: Similar to metamorphosis, the changing of an object's form, appearance, character, or function; to take ordinary images and make them uncommon, uncanny, or extraordinary
two-point perspective: A technique in art wherein construction lines representing parallel lines found in the environment appear to converge to two points positioned on the horizon line, producing the illusion of three-dimensionality on a two-dimensional surface
two-dimensional: A shape that has length and width but no height (thickness)
unity: A principle of art/design where the arrangement of elements gives the viewer a sense that all of the parts form a coherent whole, as opposed to a sense of incompletion or unrelated focal points
vanishing point: A point drawn on the horizon line to which all construction lines, representing parallel edges, appear to converge on the picture plane
variation: A change in representation of an object or person from a previous rendering, or between renderings within the same artwork; also, using a variety of materials (e.g., pencil, paint) and techniques (e.g., stipple, hatch, smudge) to create a sense of energy, emotion, or unpredictability
viewfinder: A small frame or window used by the artist to narrow the field of vision in order to more carefully focus on details or to help plan a composition for an artwork
visual pun: Drawing an object in combination with another object to create a visual play on words, or pun (e.g., sea-horse, hammer-head shark, cow-boy)
warm colours: Colours (yellow, orange, and red) that give the visual effect of advancing towards the viewer or being closer to viewer on the picture plane
warp and weft: The warp are the taut threads on a weaving loom; the weft (or woof) are the threads woven through the warp on the weaving loom
wash: A watercolour technique where the paint is thinly applied to the flat surface
wax resist: A process in which paint is applied to a surface after wax or crayon are used to draw images, and the waxed area resists the paint
weaving: The process of making cloth/containers by crossing sets of threads over and under each other

Appendix C

Grade 9 Integrated Project Sample: Golden Ratio & Proportion
(50 marks)

A. Mathematics Knowledge & Understanding (10 marks)

Define the following terms: (1 mark each)

1. Ratio:
2. Proportion:
3. Golden ratio (a.k.a. golden section, golden mean, golden cut, divine proportion, and golden proportion):
4. φ (Greek letter and numerical approximation):
5. Phidias:
6. Golden rectangle:
7. Golden triangle:
8. Fibonacci sequence:
9. Parthenon:
10. Pentagram (a.k.a. pentalpha, pentacle, and pentangle):

B. Mathematics Thinking, Inquiry, & Problem Solving (10 marks)

Solve or complete the following questions:

1. Classical aesthetics (2 marks)
In ancient Greece, the classical sculptors frequently used the navel (belly button) as the golden cut of the human body. This was evident in their sculptures of the human figure (Kouros). How do you compare to the classical standards of beauty? In other words, is your navel the golden cut of your body?

a. Measure floor to navel = ___________

b. Measure your overall height = ___________

$\frac{b}{a}$ = _______

My conclusion:

2. Golden proportion (2 marks)

Given a rectangle with dimensions 150 cm × 80 cm, determine whether it is a golden rectangle (show calculations). If not, determine what width would make it one.

3. The artist's stretcher (2 marks)

An artist wants to design a canvas that is in the shape of the golden section rectangle for her next large oil painting. If the maximum length possible for the stretcher is 240 cm, find the corresponding width (to two decimals). What will be the total perimeter of the stretcher?

4. The golden hunt (4 marks)

Analyze seven common rectangular objects around the house to determine if they are "golden."

Object	Length (b)	Width (a)	Ratio b/a	Close to φ?
Canadian $ bill				

C. Mathematics Application of Concepts (15 Marks)

Create the following items:

1. A golden section (2 marks)
On the given line AB, label the golden section (GS) with a point.

2. A golden rectangle (3 marks)
From the given square ABCD, construct and label a golden section rectangle.

3. A golden "SMARTWORK" (Secondary Math & Art Work) (10 marks)
Based on the new learning from this assignment and using the handouts given, construct a SMARTWORK that features the golden ratio in some form. This project allows for great freedom of expression. Your piece may be two- or three-dimensional. It may be narrative (tells a story), literary (using quotes or an illustrated poem), historical (perhaps a tie to math history?), ab-

stract (no recognizable objects), a collage, a sculpture—your imagination is the limit!

You may choose to simply draw a golden rectangle, or you may choose to incorporate more complex ideas from any of the golden triangle, the golden spiral, the pentagram, the lute of Pythagoras, the Fibonacci number sequence, etc.

Once again, the final SMARTWORK *must* include the golden ratio somewhere in the composition and will also be assessed for creativity and complexity.

D. Mathematics Communication (15 marks)

Use mathematical language to communicate in the following forms:

1. Written description of your "SMARTWORK" (10 marks)
Use the remainder of your page (add further sheets if necessary) to describe in detail your finished project. For example, you should explain exactly where the golden section(s) is located and the reason you chose this particular format, style, colour scheme, etc. Try to use as much mathematical language as possible (i.e., ratio, length, width, proportion).

2. Verbal presentation (5 marks)
In a brief presentation to the class (3-5 minutes), display and describe your SMARTWORK. You may wish to discuss a few points from the planning stages (show preliminary sketches/ideas?) or from the process (changes made, new ideas, etc.). Relax. Enjoy the opportunity to share your creation with others.

Selected Answers: A. Mathematics Knowledge & Understanding (10 marks)

1. Ratio: A comparison of two quantities (numbers) with the same units

2. Proportion: An equation that states that two ratios are equal
3. Golden ratio: The division of a segment into extreme and mean ratio; the name given to an irrational proportion, which has often been thought to possess some aesthetic virtue in itself, some hidden harmonic proportion in tune with the universe; a line that is divided in such a way that the smaller part is to the larger as the larger is to the whole; represented by phi = 1.61803
4. φ Greek letter and numerical approximation): phi, 1.61803
5. Phidias: Classical Greek sculptor who used the golden ratio extensively in his work; the symbol φ was chosen in his honour; created frieze panes on Parthenon (525 BC) and the huge ivory/gold statue of Athena that was housed inside the temple
6. Golden rectangle: A rectangle in which the lengths of the sides are in the golden ratio
7. Golden triangle: An isosceles triangle in which the smaller side is in golden ratio with its adjacent side
8. Fibonacci sequence: The sequence of numbers in which each term is created by adding together the two preceding terms (e.g., 1,1,2,3,5,8,13,21, . . .); the ratios of consecutive pairs converge to φ
9. Parthenon: Ancient Greek temple, located in Athens on the Acropolis, which was constructed using the golden section proportions
10. Pentagram (a.k.a. pentalpha, pentacle, and pentangle): A star-shaped figure formed by extending the sides of a regular pentagon to meet at five points.

Selected Answers: B. Mathematics Thinking, Inquiry, & Problem Solving (10 marks)

1. Classical aesthetics : When students divide their height by the distance measured from the floor to their navel, if they arrive at a number near 1.61803 (e.g., from 1.56 to 1.66), then they should indicate “golden proportions.”

2. Golden proportion:

$$\frac{150}{80} = 1.875 \neq \varphi; \therefore \text{ not golden. } \frac{150}{1.61803} = 92.7; \therefore \text{ use } 150 \times 92.7 \text{ cm}$$

3. The artist's stretcher:

$$\frac{240}{1.61803} = 148.33; \therefore\text{, golden dimensions would be } 240 \text{ x } 148.33 \text{ cm;}$$

Perimeter of Stretcher = 2(240 + 148.33) = 776.67 cm

Selected Answers: C. Mathematics Application of Concepts (15 marks)

1. A golden section: Line AB is 13.4 cm; GS label/mark should occur at 8.28 cm

$$\frac{13.4}{1.61803} = 8.28$$

2. A golden rectangle

A golden rectangle Side AB of the given square ABCD is 5.6 cm. To determine the full length of the related golden rectangle, multiply 5.6 by phi (or 1.61803), which gives 9.06 cm. Extend sides BC and AD to lengths of 9.1 cm and label these new points E and F respectively. Connect EF to form the end width of the golden rectangle whose dimensions are 5.6 cm x 9.1 cm. Diminishing or "whirling" squares can now be added by drawing a smaller square at the bottom of the rectangle CDFE (measure the sides or just subtract 9.1 – 5.6 = 3.5 cm). This is repeated in each "leftover" rectangle until the squares are too tiny to render.

Note that the golden rectangle can also be constructed, as the Greeks would have done, using only a compass and straightedge (ruler). To do this, one must first determine the midpoint of side BC, which is done by setting the compass to a length slightly larger than the estimated midpoint, and cutting arcs from both point B and then from point C. A line is drawn through the two intersections of these arcs and where it crosses line BC is midpoint M. Now, simply set the compass to width MD, position the compass at point M, and cut an arc from point D to a line extension of BC, marking the intersection point at E. Repeat steps to determine midpoint of AD (point N), and with compass positioned at N, cut an arc from point C to AD extended and labeling as point F. Connect points EF with a line to form the end width of golden rectangle ABEF.

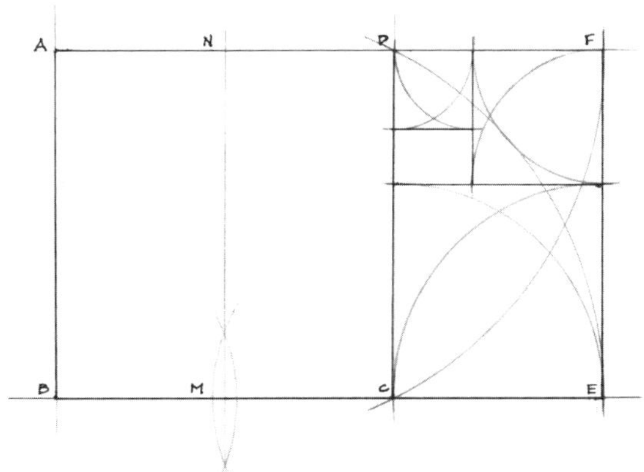

Appendix D

Grade 12 Integrated Project Sample: Fluxions & Deductions (Calculus)

(50 marks)

A. Knowledge & Understanding (10 marks)

Identify the following individuals and briefly describe their contribution(s) to the development of the Calculus (1 mark each)

1. Archimedes:
2. Fermat:
3. Barrow:

4. Newton:
5. Leibniz:
6. Bernoulli (Jacob & Johann):
7. Riemann:
8. Cauchy:

Briefly describe the two main branches of modern calculus (1 mark each)
9. Differential calculus:
10. Integral calculus:

B. Mathematics Thinking, Inquiry, & Problem Solving (10 marks)

Solve or complete the following questions:

1. Newton's legacy (3 marks)

In the winter of 1665–66, Trinity College in Cambridge closed its doors because of the Great, or Bubonic, Plague. Newton returned to his home in the English countryside. In the year that followed, Newton, at age twenty-three, made four major discoveries that would shape his world for the next three hundred years. In English history, those twelve months were commonly known as the *annus mirabilis*, the year of miracles. List the four major discoveries made by the brilliant young Newton:

a.
b.
c.
d.
e. Name Newton's most famous work, published in 1687, which presented both his astronomical and dynamical discoveries in one complex theory of the universe:

2. Newton's method (3 marks)

This method is used to approximate roots of equations for situations where standard techniques are either too complicated or non-existent. It is called an iterative method, wherein one successively calculates the terms of a sequence of approximations to the solution. The equation for x_{n+1}, which relates successive terms in the sequence, is called a recursion formula, and the one for Newton's Method is the following:

$$x_{n+1} = x_n - \frac{f(x_n)}{f'(x_n)}, \quad n = 1, 2, \ldots$$

with x_1 suitably chosen.

Use Newton's method to approximate a root of $f(x) = x^3 - 2x - 5 = 0$, starting with $x_1 = 2$ (to 5 decimal places).

3. Newton's Law of Cooling (4 marks)

Newton's Law of Cooling states that the rate at which a body cools to the temperature of its surroundings is proportional to the temperature difference between the body and its surroundings.

The formula used is $T = A + (T_o - A)\, e^{kt}$, in which A is the temperature of the surroundings, T_o is the initial temperature of the body, and k is a characteristic of the material of the body.

Solve the following problem using Newton's Law of Cooling:

The temperature of a cup of coffee is 80°C when brought into a room where the temperature is 22°C. After 10 minutes, the temperature of the coffee is 60°C. How long will it take for the temperature to reach 30°C?

C. Mathematics Application of Concepts (20 marks)

> *I do not know what I may appear to be to the world, but to myself I seem to have been only like a boy playing on the sea-shore, and diverting myself in now and then finding a smoother pebble or a prettier shell than ordinary, whilst the great ocean of truth lay all undiscovered before me.* Sir Isaac Newton (1642–1727)

> *I would walk twenty miles to my worst enemy if I could learn something from it.* Gottfried Wilhelm Leibniz (1646–1716)

A Baroque "SMARTWORK" (Secondary Math & Art Work) (20 marks)

The exciting historical period from1600–1750, often referred to in music and visual arts as the Baroque, featured monumental achievements and advances in all the disciplines. It was the time of Bach, Vivaldi, Handel, and Scarlatti, who revolutionized the musical world. In literature, authors such as Shakespeare, Marlowe, Milton, and Jonson were redefining poetry and prose. In the world of art, Caravaggio, Zurbaran, Poussin, Rubens, Vermeer, and Rembrandt were painting in a powerful new genre involving a sense of movement, light/dark contrasts, infinite space, high realism, and a deep sense of spirituality. The Roman Catholic church was a highly influential patron of the arts, as was its Counter Reformation, employing the arts as a means of spreading the faith. And finally, scientific and mathematical minds such as Galileo, Barrow, Newton, and Leibniz were making extraordinary scientific and mathematical discoveries that would have significant impact on the centuries to follow.

Based on the new learning from this assignment, particularly with respect to the contributions made by Sir Isaac Newton, design a Baroque SMARTWORK that juxtaposes a selection of ideas from this period that interest you the most. This project may take any form, using any media, and is thereby wide open to personal interpretation. Some ideas might include a poster, a sculpture, a collage, a mobile, a mathematical or physical model or set of questions with diagrams, a song, a dramatic narrative, a costume, etc. Your imagination is the limit. Push this limit. Develop something uniquely original to you, as you absorb all the richness and explosive energy of the Baroque era. There should be an obvious tie to the Calculus somewhere within the overall design of your SMARTWORK. Good luck and go for Baroque.

D. Mathematics Communication (10 marks)

Use mathematical language to communicate in the following forms:

1. Written description of your "SMARTWORK" (5 marks)

Use the remainder of your page (add further sheets if necessary) to describe in detail your finished project. For example, you should try to use as much mathematical language as possible (differentiation, laws, methods, integration, fluxions, etc.) wherever appropriate in explaining your final design.

2. Verbal presentation (5 marks)

In a brief presentation to the class (2 to 3 minutes), display and describe your SMARTWORK. You may wish to discuss a few points from the planning stages (show preliminary sketches/ideas?) or from the process (changes made, new ideas, etc.). Relax. Enjoy the opportunity to share your creation with others.

Select Project Solutions for Fluxions & Deductions

Answers for Part A. Knowledge & Understanding

1. Archimedes: (287–212 BC) Greek mathematician, physicist, engineer, inventor, and astronomer; generally considered to be the greatest mathematician of antiquity, and one of the greatest of all time
2. Pierre de Fermat: (1601–1665) A French lawyer and amateur mathematician who made notable contributions to analytic geometry, probability, and optics; best known for Fermat's Last Theorem
3. Isaac Barrow: (1630–1677) An English Christian theologian and mathematician who is generally given credit for the discovery of the fundamental theorem of calculus; teacher of Isaac Newton
4. Sir Isaac Newton: (1643–1727) An English physicist, mathematician, astronomer, natural philosopher, alchemist, and theologian; considered among the greatest and most influential scientists of all time; author of *Mathematical Principles of Natural Philosophy* (1687), which laid the foundations for classical mechanics
5. Gottfried Leibniz: (1646–1716) A German philosopher, mathematician, and logician who is probably most well known for having invented the differential and integral calculus (independently of Sir Isaac Newton); wrote broadly in the areas of mathematics, logic, science, history, law, and theology

6a. Jacob Bernoulli: (1654–1705) A Swiss mathematician, astronomer, and theologian; first to solve separable differential equations; founded a school for mathematics and the sciences in Basel (1682)

6b. Johann Bernoulli: (1667–1748) A Swiss mathematician who was one of the many prominent mathematicians in the Bernoulli family (younger brother to Jacob); known for his contributions to infinitesimal calculus; teacher of the young Leonhard Euler, another famous mathematician

7. Georg Riemann: (1826–1866) A German mathematician who made lasting contributions to analysis and differential geometry; his published works opened up research areas combining analysis with geometry
8. Cauchy: (1789–1857) A French mathematician who started the project of formulating and proving the theorems of infinitesimal calculus in a rigorous manner; a profound mathematician whose writings cover the entire range of mathematics and mathematical physics
9. Differential calculus: A subfield of calculus concerned with the study of the rates at which quantities change; one of the two traditional divisions of calculus, the other being integral calculus
10. Integral calculus: A branch of mathematics concerned with the study of the determination, properties, and application of integrals (a function which satisfies a given differential equation); one of the two traditional divisions of calculus, the other being differential calculus

Answers for Part B. Mathematics Thinking, Inquiry, & Problem Solving (10 marks)

1. Newton's legacy

a. Law of Universal Gravitation
b. Diffraction of colours, using a prism, into the visible colour spectrum
c. Binomial series for negative and fractional exponents
d. Differential and integral calculus (what he called "fluxions")
e. Famous 1687 work: *Philosophiae Naturalis Principia Mathematica (Mathematical Principles of Natural Philosophy)*

2. Newton's method

$$x_{n+1} = x_n - \frac{f(x_n)}{f'(x_n)}, \quad n = 1, 2, \ldots$$

$$f(x) = x^3 - 2x - 5 = 0, \text{ starting with } x_1 = 2$$

$$f'(x) = 3x^2 - 2$$

$$x_{n+1} = x_n - \left[\frac{x_n^3 - 2x_n - 5}{3x_n^2 - 2}\right] = 2 - \frac{8 - 4 - 5}{12 - 2} = 2 - \frac{1}{10} = 2.1$$

$$x_3 = 2.1 - \left[\frac{(2.1)^3 - 2(2.1) - 5}{3(2.1)^3 - 2}\right] \cong 2.094568$$

$$x_4 \cong 2.094551482$$

$$x_5 \cong 2.094551482$$

$$\therefore\ x = 2.09455$$

3. Newton's Law of Cooling

$$T = A + (T_0 - A)e^{kt}$$

Solve for the variable t, where $T_0 = 80$, $A = 22$, $\mathrm{T}_{(10)} = 60$, $\mathrm{T}_{(t)} = 30$

$$60 = 22 + (80\text{-}22)e^{k10}$$

$$60 = 22 + (58)e^{k10}$$

$$38 = (58)e^{k10}$$

$$\frac{38}{58} = (e^k)^{10}$$

$$ln\ \frac{38}{58} = 10 ln e^k$$

$$k = \frac{ln\frac{38}{58}}{10}$$

$$\mathbf{30} = 22 + (80 - 22)e^{\left(\frac{ln\frac{38}{58}}{10}\right)t}$$

$$ln\frac{8}{58} = t\left(\frac{ln\frac{38}{58}}{10}\right)$$

$$t = \frac{10^{ln\frac{8}{58}}}{ln\frac{38}{58}}$$

t = 46.848 minutes ≅ 46 minutes, 51 seconds.

∴ it would take approximately 46.8 minutes for the coffee to cool to this temperature.

Appendix E

Other Recommended Print and Online Resources

Abbott, E. A. (1998/1884). *Flatland: A romance of many dimensions.* New York, NY: Penguin.

Abbott, J. (1997). To be intelligent. *Educational Leadership, 54*(6), 6–10.

Adler, B. (1998, September). You gotta have art. *Instructor,* 86–87.

Alesandrini, J. L. (2000). Integrating the arts: A nationwide survey of current practices in K–8 visual art education. *Dissertation Abstracts International, 60* (08), 2771.

Ames, L. (1982). *The dot, line and shape connection, or how to be driven to abstraction.* New York, NY: Doubleday.

Anderson, T. (2002). Mandala: Constructing peace through art. *Art Education, 55*(3), 33–39.

Aoki, T. (1991). *Inspiriting the curriculum and pedagogy: Talks to teachers.* Edmonton, AB: University of Alberta.

Arguelles, J., & Arguelles, M. (1995). *Mandala.* Boston, MA: Shambhala Publications.

Armstrong, T. (1994). Multiple intelligences: Seven ways to approach curriculum. *Educational Leadership, 52*(3), 26–28.

Ascher, M. (1991) *Ethnomathematics: The multicultural view of mathematical ideas.* New York, NY: Springer.

Aston, M. (2000). *The panorama of the Renaissance.* New York, NY: Abradale Press.

Auvil, K. W. (1997). *Perspective drawing* (2nd ed.). Mountain View, CA: Mayfield.

Bayer, H., Gropius W., & Gropius I. (Eds.). (1959). *Bauhaus 1919-1928.* Boston, MA: Charles T. Branford.

Beane, J. A. (1995). Curriculum integration and the disciplines of knowledge. *Phi Delta Kappan, 76*(8), 616–622.

Beckmann, P. (1970). *The history of Pi.* Boulder, CO: Golem Press.

Bergland, D. (1996). The digital image: On the frontiers of a new art education. *Canadian Review of Art Education, 25*(4), 141–152.

Berlinski, D. (1995). *A tour of the Calculus.* New York, NY: Pantheon Books.

Bezuszka, S., Kenney, M., & Silvey, L. (1977). *Tessellations: The geometry of patterns.* Palo Alto, CA: Creative Publications.

Biller, J. (1995). *Math in art or art in math.* Paper presented at the Annual National Conference on
Liberal Arts and Education of Artists, New York, NY.

Blatner, D. (1997). *The joy of Pi.* New York, NY: Walker and Company.

Boles, M., & Newman, R. (1992). *The surface plane: The golden relationship in art, math and nature.*
Bradford, MA: Pythagorean Press.

Bouleau, C. (1980). *The painter's secret geometry: A study of composition in art.* New York, NY: Hacker Art Books.

Bouwsma, W. (1972). *Geometry for teachers.* New York, NY: MacMillan.
Boyer, C. B. (1991). *A history of mathematics* (2nd Ed.). Toronto, ON: John Wiley & Sons.
Braxton, B., Gonsalves, P., Lipner, L., & Barker, J. (1995). *Math around the world.* Berkeley, CA: University of California at Berkeley.
Brewer, E. J. (1999). Geometry and Op art. *Teaching Children Mathematics, 6*(4), 220–236.
Brewer, T. (2002). An examination of intrinsic and instrumental instruction in art education. *Studies in Art Education, 43*(4), 354–372.
Brommer, G. (1978). *The art of collage.* Worcester, MA: Davis.
Brophy, J., & Alleman, J. (1991). A caveat: Curriculum integration isn't always a good idea. *Educational Leadership, 49*, 66–67.
Brownell, S. (1993). The flower as mandala. *School Arts, 93*(1), 35.
Burnet, N., & Matsen, B. (2002). *The shape of life.* Monterey, CA: Monterey Bay Aquarium Press.
Buxton, L. (1991). *Math panic.* Portsmouth, NH: Heinemann.
Camp, D. (2000). Benoit Mandelbrot: The Euclid of fractal geometry. *The Mathematics Teacher, 93*(8), 708–712.
Campbell, B., Campbell, L., & Dickinson, D. (1992). *Teaching and learning through multiple intelligences.* Stanwood, WA: New Horizons for Learning.
Case, R. (1992). The anatomy of curricular integration. *The Recorder, 35*(1), 30–36.
Chatterley, L. J., & Peck, D. M. (1995). We're crippling our kids with kindness! *Journal of Mathematical Behavior, 14*, 429–436.
Cheatham, F., Cheatham, J., & Owens, S. (1987). *Design concepts and applications* (2nd ed.). Englewood Cliffs, NJ: Prentice-Hall.
Closs, M. (1986). *Native American mathematics.* Austin, TX: University of Texas Press.
Cornell, C. (1999). "I hate math! I couldn't learn it, and I can't teach it!" *Childhood Education, 75*(4), 225–231.
Cornell, J. (1994). *Mandala.* Wheaton, IL: Quest Books.
Corrain, L. (1997). *Masters of art: The art of the Renaissance.* Chicago, IL: Peter Bedrick Books.
Dantzic, C. (1990). *Design dimensions: An introduction to the visual surface.* Englewood Cliffs, NJ: Prentice-Hall.
Deitch, J., Halley, P., Friedman, D., & Steinbach, H. (1988). *Cultural geometry.* Athens, Greece: Deste Foundation for Contemporary Art.
Devlin, K. (1994). *Mathematics: The science of patterns: The search for order in life, mind and the universe.* New York, NY: Scientific America Library.
Devlin, K. (1998). *Life by the numbers.* New York, NY: John Wiley & Sons.
Dewey, J. (1934). *Art as experience.* New York, NY: Minton, Balch, & Co.
Dewey, J. (1973). *Experience and education:* New York, NY: Collier Books.
Dubosque, D. (1999). *Draw 3-D: A step-by-step guide to perspective drawing.* Peel Productions. Retrieved from http://drawbooks.com/peelbooks.com/index.html.
Edwards, B. (1986). *Drawing on the artist within.* New York, NY: Simon & Schuster.
Eisner, E. (1972). *Educating artistic vision.* New York, NY: Macmillan.
Eisner, E. (1983). The art and craft of teaching. *Educational Leadership, 40*(4), 4–13.

Eisner, E. (1987). *The role of discipline-based art education in America's schools.* (ERIC Document Reproduction Service No. ED 298034)

Ellison, E. (1972). *Fun with lines and curves.* New York, NY: Lothrop, Lee & Shepard.

Emmer, M. (1993). *The visual mind: Art and mathematics.* Cambridge, MA: MIT Press.

Escher, M. C. (1989). *Escher on Escher: Exploring the infinite.* New York, NY: H. N. Abrams.

Fein, S. (1993). *First drawings: Genesis of visual thinking.* Pleasant Hill, CA: Exelrod Press.

Feldman, E. (1970). *Becoming human through art.* Englewood Cliffs, NJ: Prentice Hall.

Field, J. V. (1997). *The invention of infinity mathematics and art in the Renaissance.* Oxford, UK: Oxford University Press.

Fincher, S. (1991). *Creating mandalas.* Boston, MA: Shambhala Publications.

Fogarty, R. (1999). Architects of the intellect. *Educational Leadership, 57*(3), 76–78.

Fogarty, R., & Stoehr, J. (1995). *Integrating curricula with multiple intelligences: Teams, themes, and threads.* Pailtine, IL: IRI/Skylight Publishing.

Frayling, C., Frayling, H., & Van der Meer, R. (1992). *The art pack.* New York, NY: Alfred A. Knopf Publishing.

Gardiner, M., Fox, A., Knowles, F., & Jeffrey, D. (1996). Arts improve reading and math. *Teaching Music, 4*(2), 8.

Gardner, H., & Boix-Mansilla, V. (1994). Teaching for understanding within and across the disciplines. *Educational Leadership, 51*, 14–18.

Gatto, J., Porter, A., & Selleck, J. (1987). *Exploring visual design.* Worcester, MA: Davis Publications.

Gerdes, P. (1999). *Geometry from Africa: Mathematical and educational explorations.* Washington, D.C.: The Mathematical Association of America.

Ghyka, M. (1977). *The geometry of art and life.* New York, NY: Dover.

Goldberg, M. (2001). *Arts and learning: An integrated approach to teaching and learning in multicultural and multilingual settings* (2nd ed.). New York, NY: Longman.

Goldstein, E. (1987). *Creating and understanding art.* Champaign, IL: Garrard Publishing Company.

Granger, T. (2000). Math is art. *Teaching Children Mathematics. 7*(1), 10.

Gregory, R., Gregory, L., & Gombrich, E. (Eds). (1973). *Illusion in nature and art.* New York, NY: Charles Scribner's Sons.

Hofstadter, D. R. (1979). *Godel, Escher, Bach: An eternal golden braid.* New York, NY: Basic Books.

Huntley, H. E. (1970). *The divine proportion: A study in mathematical beauty.* New York, NY: Dover.

Ivins, W. M., Jr. (1946). *Art and geometry: A study in space intuitions.* Mineola, NY: Dover Publications.

Jacobs, H. H. (1991). Planning for curriculum integration. *Educational Leadership, 49*, 27–28.

Johnson, M. (1990). *Teach your child to draw: Bringing out your child's talents and appreciation for art.* Los Angeles, CA: Lowell House.

Kemp, M. (1990). *The science of art: Optical themes in western art from Brunelleschi to Seurat.* London, UK: Yale University Press.

King, A. (1999). *Math for fun projects.* Brookfield, CT: Copper Beech Books.

Kleiman, G. (1991). Mathematics across the curriculum. *Educational Leadership, 49*(2), 48–51.
Lawlor, R. (1982). *Sacred geometry: Philosophy and practice*. New York, NY: Thames & Hudson.
Le Corbusier. (1954). *Le Modular: The harmonious measure to the human scale universally applicable to architecture and mechanics* (P. de Francia & A. Bostock, Trans.). Cambridge, MA: MIT Press.
Linn, C. (1974). *The golden mean, mathematics and the fine arts*. Garden City, NY: Doubleday.
List, L. (1982). *Music, art, and drama experiences for the elementary curriculum*. New York, NY: Teachers College Press.
Livio, M. (2005). *The equation that couldn't be solved: How mathematical genius discovered the language of symmetry*. New York, NY: Simon & Schuster.
Locher, J. L., & Thé, E. (2000). *The magic of M. C. Escher*. New York, NY: H. N. Abrams.
Lovejoy, M. (1997). *Postmodern currents: Art and artists in the age of electronic media* (2nd ed.). Upper Saddle River, NJ: Prentice Hall.
Mankiewicz, R. (2000). *The story of mathematics*. London, UK: Cassell.
March, L., & Steadman, P. (1971). *The geometry of environment: An introduction to special organization in design*. Cambridge, MA: MIT Press.
Miller, C., & Veenstr, T. (2002). Fibonacci: Beautiful patterns, beautiful mathematics. *Mathematics Teaching in the Middle School, 7*(5), 298–305.
Mlodinow, L. (2001). *Euclid's window: The story of geometry from parallel lines to hyperspace*. New York, NY: The Free Press.
Monk-Kidd, S. (2002). *The secret life of bees*. New York, NY: Penguin Books.
Morgan, H. (1995). Integrating art in the elementary classroom. *Journal of the Canadian Society for Education Through Art, 26*(1), 12–14.
Myers, J. F. (1989). *The language of visual art: Perception as a basis for design*. Toronto, ON: Harcourt.
Olson, A. (1975). *Mathematics through paper folding*. Reston, VA: National Council of Teachers of Mathematics.
Patterson, B. (2001). Blending art and geometry with precision. *Arts and Activities, 130*(1), 46.
Paulos, J. (1992). *Beyond numeracy: Ruminations of a numbers man*. New York, NY: Vintage Books.
Paulos, J. A. (1988). *Innumeracy: Mathematical illiteracy and its consequences*. New York, NY: Hill & Wang.
Pedoe, D. (1958). *The gentle art of mathematics*. New York, NY: Dover.
Piaget, J. (1972). Intellectual development from adolescence to adulthood. *Human Development, 15*(1), 1–12.
Piaget, J., Inhelder, B., & Szeminska, A. (1964). *The child's conception of geometry* (E. A. Lunzer, Trans.). New York, NY: Harper & Row.
Ritter, D. (1995). *Math art: Featuring 33 innovative art projects*. Cypress, CA: Creative Teaching Press.
Roukes, N. (1988). *Design synectics: Stimulating creativity in design*. Worcester, MA: Davis Publications.
Schwartz, A. (2000). Axing math anxiety. *The Education Digest, 65*(5), 62–64.

Seymour, D., & Britton, J. (1989). *Introduction to tessellations*. Palo Alto, CA: Dale Seymour Publications.
Smith, R. (1995). *An introduction to perspective*. London, UK: Dorling Kindersley.
Stewart, I. (1998). *Life's other secret: The new mathematics of the living world*. New York, NY: John Wiley & Sons.
Stewart, I. (2001). *Flatterland: Like Flatland, only more so*. Cambridge, UK: Perseus.
Strother, Z., & Hailemariam, S. (2000, Spring). Art that heals: The image as medicine in Ethiopia. *African Arts, 33*(1), 12–15.
Sylwester, R. (1998). Art for the brain's sake. *Educational Leadership, 56*(3), 31–33.
Taylor, A. (1974). *Math in art*. Hayward, CA: Activity Resources Company.
Thompson, K. (1995). Maintaining artistic integrity in an interdisciplinary setting. *Art Education, 44*(20), 38–45.
Vacc, N. N. (1999). Exploring fractal geometry with children. *School Science and Mathematics, 99*(2), 77–83.
Valens, E. G. (1964). *The number of things: Pythagoras, geometry and humming strings*. Toronto, ON: Clarke, Irwin & Co.
Van Cleave, J. (1994). *Geometry for every kid: Easy activities that make learning geometry fun*. New York, NY: John Wiley & Sons.
Van der Meer, R., & Gardner, B. (1994). *The math kit: A three-dimensional tour through mathematics*. New York, NY: Macmillan.
Vars, G. (1991). Integrated curriculum in historical perspective. *Educational Leadership, 49*(2), 14–15.
Watson-Newlin, K. (1995, October). Stairways to integrated learning. *School Arts*, 43–44.
West, K. (1995). *Basic perspective for artists: A guide to the creative use of perspective in drawing, painting and design*. New York, NY: Watson-Guptill.
Whitehead, A. (1929). *Aims of education*. New York, NY: Free Press.
Wilman, C. (1966). *Seeing and perceiving*. New York, NY: Pergamon Press.
Woodman, A., & Albany, E. (1992). *Mathematics through art and design* (Rev. ed.). London, UK: HarperCollins.
Wright, L. (1983). *Perspective in perspective*. London, UK: Routledge and Kegan Paul.
Zaslavsky, C. (1998). *Math games and activities from around the world*. Chicago, IL: Chicago Review Press.

Other Recommended Online Resources

Art + Math = X (Farsi)
http://plus.maths.org/content/os/issue37/features/farsi/index
ARTeFACTS (Jarvis)
http://faculty.nipissingu.ca/danj/other/ARTeFACTS.htm
Bathsheba Grossman Sculpture
http://www.bathsheba.com/

Bridges Organization: Math Connections in Art, Music, & Science
http://www.bridgesmathart.org/
GeoGebra Mathematics Education Software
http://www.geogebra.org/cms/
Google Art Project
http://www.googleartproject.com/
(See below for a description of how to use this great art resource.)
Google Earth
http://www.google.com/earth
Helaman Ferguson Sculpture
http://www.helasculpt.com/
Journal of Mathematics and the Arts
http://www.informaworld.com/smpp/title~db=all~content=t755420531~tab= issueslist
Learning Through the Arts (Royal Conservatory of Music, Canada)
http://www.ltta.ca/
Math Art Fun
http://mathartfun.com/
Mathematical Imagery (AMS)
http://www.ams.org/mathimagery/
MATHeMATRIX (Jarvis)
http://www.nipissingu.ca/faculty/danj/other/MATHeMATRIX.htm
Mathematics and its Connections to the Arts and Sciences (MACAS 2)
http://www.macas2.sdu.dk/
Museum of Harmony and Golden Section (Stakhov)
http://www.goldenmuseum.com/index_engl.html
SET Game of Visual Perception (Falco)
http://www.setgame.com/set/index.html
Tessellations.org (Bareiss)
http://www.tessellations.org

The Google Art Project

Launched in February 2011, The Google Art Project (GAP) originally involved seventeen major world galleries. After a major expansion in 2012, the GAP now includes over 30,000 works of art (sculpture, architecture, drawing, paintings) from over 150 collections from forty participating countries. Google Art is now a unique online art experience. Users can explore a wide range of artworks at brushstroke-level detail, take a virtual tour of a museum, and even build their own collections to share.

For example, to view Goya's painting, *The Third of May*, open Google Earth (free software downloadable from the Internet), check off the box for "3-D Buildings" in the left-hand sidebar preferences, search for Prado Museum in the search box, and zoom in to this museum, which is located in Madrid, Spain. Click on the museum, and a pop-up window displays all of the artwork images available for viewing. Select Goya's *The Third of May* and you will see an ultra-high resolution image of the painting. The Google Art Project is an excellent example of how art (images, museums) and mathematics (i.e., technical aspects of lighting, filming, storing, accessing, displaying, and personalizing imagery) continue to coincide and synergize in the digital era.

Index of Learning Experiences

Chapter 3

Observing and Documenting Plant Growth /59
Printmaking and Wax Rubbings with Plants /60
Observing and Documenting Plant Characteristics /63
Exploring and Creating Spiral Patterns /66
Creating Fractals /68
Exploring Positive and Negative Shapes with Plants /69
Observing, Drawing, and Reinterpreting the Flower /70
Create Macro Drawings with a Viewfinder /71
Flower Symbolism and Cultural Significance /72
Drawing and Distorting Plants Using Grids /74
Ferocious Flowers /74
Exploring Bilateral Symmetry with Plant Shapes /76
Exploring Geometric Shapes in Plants and Flowers /77
Kaleidoscopic Designs /77
Sketchbook Assignments Involving Trees /79

Measuring Tree Growth /82
Observing and Creating Tree Designs /84
Exploring Leaves /86
Creating Landscapes /89
Exploring Rainbows /93
Rainbows, Culture, and Illustrated Creative Writing /94
Observing and Depicting Cloud Formations /94
Investigating the Natural Environment in Literature /96

Chapter 4

Sketchbook Assignments /100
Animal Portraits Using Measurements and Comparison /104
Grid Drawings Using Dilatation (Scaling) /105
Animal Portraits Using a Coordinate System /108
Mystery Animal Plot /109
Animals in Art and Design /109
Camouflage in Artwork /111
Wire Sculpture Figure in Motion /112
Tessellation Investigation /114
Tessellations Involving Fauna Forms /115
Creating a Dossier of Aquatic Imagery, Text, and Video 118
Drawing Dinosaurs and Sculpting Skeletons /120
Creating an Original Myth/Legend Creature and Narrative /121

Chapter 5

Human Body Measurement /125
Measuring and Graphing Human Body Parts /125
Human Body Ratios/ 127
Who Fits in a Square? /128
Exploring Scatter Plots, Linear Regression, and Lines of Best Fit /129

The Golden Ratio in the Human Body /132
Profile Shadow and Reflection /133
The Fingers and the Spaces /135
Fractions in the Human Hand /137
Contour Drawings of the Human Hand /138
Drawing the Human Eye /139
Ratio and Proportions in the Human Foot /142
Exploring the Human Ear /143
Body Part Prints /144
Portrait Drawing Using Proportions and Estimation /146
Portrait Drawing Using Direct Measurement Transfer /147
Drawing and Scaling the Human Form Using Grids /149
Portrait Reflections /151
Silhouette Portraits /152
Gesture Drawing /153
Blind Contour and Mass Drawings of the Body /155
The Human Body Made Geometric /155
Portraits in Caricature and with Distortion /156
Wire Sculpture and Action Figures /157
Papier-Mâché and Fabric Soft Sculptures /158
Caricature Faces /159
Creating People Boxes /160
The Human Body and Metamorphosis /160
Drawing a Sasquatch /161
Leprechauns and Proportion /162

Chapter 6

Architecture Field Trip and Space Design /173
Indigenous Architecture Research and Design /178
Architectural Themes and Cross-Disciplinary Research /181

Drawing Buildings and Landscapes /183
Drawing a Building Using Coordinates /186
One-Point Perspective Drawing /189
Two-Point Perspective Drawing /190
Exploring Illusion of Depth Techniques /191
Form versus Function Analysis of Structures /193
Build Your Dream Home /194
Rendering Interiors with Perspective /196
Invented Landscapes and Fantasy Architecture /198
Exploring Bridges /199

Chapter 7

Measuring and Observing a Room Interior /204
Form and Function in Design /205
Drawing Interiors /206
Diorama Creation /208
Still Life /208
Exploring Patterns in the Environment /209
Creating a Golden Rectangle and Triangle /212
Tiled Floor and Bauhaus Objects /213
Designing Logos, Trademarks, Medals, and Flags /214
Name and Numbers in Perspective /214
Creating 2-D Nets for 3-D Solids /216
Designing Product Packaging /216
Methods of Making Circles /219
Measuring Distance with a Trundle Wheel /220
The Potter's Wheel /221
Creating a Mandala /223
Drawing an Accurate Easter Egg /224
Currency Design /225

Calendar Design /227
Time and Tide Artwork /227
Creating Sundials and Model Clock Towers /230
Game Design and Observation /232
Video Documentary Creation /233

Image Credits

D. Jarvis: Daniel Jarvis, Nipissing University, North Bay, Ontario
I. Naested: Irene Naested, Mount Royal University, Calgary, Alberta
MRU student work: Mount Royal University student work, used with permission, courtesy Irene Naested, Calgary, Alberta
Shutterstock: *www.shutterstock.com*

The authors have made their best efforts to properly credit all images. Please contact us with any errors or deficiencies and we will make appropriate corrections and additions in future editions.

CHAPTER 1: C1-1 D. Jarvis; C1-2 I. Naested; C1-3 I. Naested; C1-4 Shutterstock: An Indian Teepee Set Up in a Meadow among Pine Woods, Image ID: 57261730, Copyright: Benedictus; C1-5 I. Naested; C1-6 Shutterstock: Parthenon on the Acropolis of Athens. Greece, Image ID: 82605838, Copyright: Uleanov; C1-7 Shutterstock: Fresco in Rafael Rooms, Vatican Museums, Image ID: 51885448, Copyright: Bojan Pavlukovic; C1-8 I. Naested; C1-9 D. Jarvis; C1-10A, C1-10B, and C1-10C D. Jarvis

CHAPTER 2: C2-1 D. Jarvis, C2-2 I. Naested; C2-3 MRU student work; C2-4 MRU student work; C2-5 MRU student work; C2-6 MRU student work; C2-7 MRU student work; C2-8 MRU student work; C2-9 I. Naested; C2-10 Isabelle Hunt-Johnson, watercolour, used with permission; C2-11 I. Naested; C2-12 *Thumbnail Sketches*, mixed media, D. Jarvis, 1999; C2-13; C2-14 *Amgine Raw,* acrylic, oil, and gold leaf on masonite, D. Jarvis, 1995; C2-15 I. Naested; C2-16 I. Naested; C2-17 I. Naested; C2-18A and C2-18B D. Jarvis; C2-19 Shutterstock: Boy Holding Binoculars, Image ID: 99256619, Copyright: Sergii Figurnyi

CHAPTER 3: C3-1 D. Jarvis; C3-2 Shutterstock: Corn Isolated on White, Image ID: 69705574, Copyright: Africa Studio; C3-3 D. Jarvis; C3-4 D. Jarvis; C3-5 I. Naested; C3-6 D. Jarvis; C3-7 I. Naested; C3-8 Shutterstock: Romanesco Cabbage, Image ID: 19689667, Copyright: Nataliya Hora; C3-9 I. Naested; C3-10 Shutterstock: Nautilus Shell Interior on Black,

Image ID: 90930266, Copyright: Christian Delbert; C3-11 Shutterstock: Sunflower Close-Up, Image ID: 88882021, Copyright: Demetrio Mancini; C3-12 I. Naested; C3-13 I. Naested; C3-14 Shutterstock: Green Fern Leaf, Image ID: 31784434, Copyright: Kostyantyn Ivanyshen; C3-15 D. Jarvis; C3-16 Meagan Bothan and Carley Foisy; C3-17 *Metamorphosis Drawing of Swiss Army Knife into Lada Car*, graphite on paper, D. Jarvis, 1995; C3-18 I. Naested; C3-19 D. Jarvis; C3-20 D. Jarvis; C3-21 D. Jarvis; C3-22 I. Naested; C3-23 I. Naested, 2001; C3-24 I. Naested; C3-25 MRU student work; C3-26 MRU student work; C3-27 Shutterstock: Worker Bees on Honeycomb, Image ID: 56105794, Copyright: LilKar; C3-28 D. Jarvis; C3-29 MRU student work; C3-30 D. Jarvis; C3-31 MRU student work; C3-32 MRU student work; C3-33 MRU student work; C3-34 D. Jarvis; C3-35 I. Naested; C3-36 Shutterstock: Vector Silhouette Pipe Smoker (Sherlock), Image ID: 70800004, Copyright: Dimec; C3-37 Shutterstock: Web in Dew Drops on Dark Background, Image ID: 53777113, Copyright: Milena; C3-38 I. Naested; C3-39 I. Naested; C3-40 Shutterstock: Pine Cone on Black Background, Image ID: 91365149, Copyright: alexsvirid; C3-41 I. Naested, 2010; C3-42 I. Naested; C3-43 D. Jarvis; C3-44 *Empire,* acrylic, oil, and gold leaf on masonite, D. Jarvis, 1995; C3-45 D. Jarvis

CHAPTER 4: C4-1 Shutterstock: White Tiger, Image ID: 28671337, Copyright: Oleg Kozlov; C4-2 D. Jarvis; C4-3 I. Naested; C4-4 MRU student work; C4-5 MRU student work; C4-6 MRU student work; C4-7 MRU student work; C4-8 I. Naested; C4-9 D. Jarvis; C4-10 MRU student work; C4-11 MRU student work; C4-12 MRU student work; C4-13 MRU student work; C4-14 MRU student work; C4-15 MRU student work; C4-16 MRU student work; C4-17 I. Naested; C4-18 MRU student work; C4-19 MRU student work; C4-20 MRU student work; C4-21 D. Jarvis; C4-22 GeoGebra, courtesy D. Jarvis; C4-23 D. Jarvis; C4-24 MRU student work; C4-25 MRU student work; C4-26 D. Jarvis; C4-27 D. Jarvis; C4-28 D. Jarvis; C4-29 D. Jarvis; C4-30 D. Jarvis; C4-31 D. Jarvis; C4-32 MRU student work; C4-33 MRU student work; C4-34 MRU student work; C4-35 MRU student work; C4-36 MRU student work; C4-37 MRU student work; C4-38 MRU student work; C4-39 MRU student work; C4-40 Shutterstock: Dino Head, Image ID: 17658697, Copyright: restless; C4-41 MRU student work; C4-42 MRU student work

CHAPTER 5: C5-1 I. Naested; C5-2 I. Naested; C5-3 D. Jarvis and I. Naested; C5-4 D. Jarvis; C5-5 D. Jarvis; C5-6 D. Jarvis; C5-7 D. Jarvis; C5-8 Shutterstock: X-ray Human Hand, Image ID: 75247996, Copyright: antoniomas; C5-9 I. Naested; C5-10 D. Jarvis; C5-11 I. Naested; C5-12 D. Jarvis; C5-13 D. Jarvis; C5-14 Shutterstock: 30 Detailed Fingerprints, Image ID: 72073594, Copyright: Undergroundarts.co.uk; C5-15 I. Naested; C5-16 I. Naested; C5-17 D. Jarvis; C5-18 I. Naested; C5-19 I. Naested; C5-20 I. Naested; C5-21 I. Naested; C5-22 D. Jarvis; C5-23 MRU student work; C5-24 D. Jarvis; C5-25 MRU student work; C5-26 MRU student work

CHAPTER 6: C6-1 D. Jarvis, C6-2 D. Jarvis; C6-3 D. Jarvis; C6-4 D. Jarvis; C6-5 D. Jarvis; C6-6 D. Jarvis; C6-7 D. Jarvis; C6-8 Shutterstock: Aerial of the Pentagon, the Department of Defense Headquarters in Arlington, Virginia, Image ID: 35841916, Copyright:

Frontpage; C6-9 D. Jarvis; C6-10 D. Jarvis; C6-11 D. Jarvis; C6-12 Shutterstock: Igloo in a Winter Mountain Setting, Image ID: 41074081, Copyright: Tyler Olson; C6-13 Shutterstock: The Dome of the Rock, Jerusalem, Israel, Image ID: 70304869, Copyright: J van der Wolf; C6-14 Shutterstock: Pyramids in Giza, Image ID: 30216103, Copyright: sculpies; C6-15 D. Jarvis; C6-16 Shutterstock: 3 Columns in Different Styles: Ionic, Doric, Corinthian, Image ID: 93841513, Copyright: mart; C6-17 I. Naested; C6-18 D. Jarvis; C6-19 Shutterstock: Facade of the Jewish Museum in Berlin (Germany), Image ID: 26452918, Copyright: erzetic; C6-20 MRU student work; C6-21 Calgary Science School student work, used with permission; C6-22 I. Naested; C6-23 MRU student work; C6-24 D. Jarvis; C6-25 MRU student work; C6-26 I. Naested; C6-27 Calgary Science School student work, used with permission; C6-28 I. Naested; C6-29 D. Jarvis; C6-30 D. Jarvis

CHAPTER 7: C7-1 I. Naested; C7-2 I. Naested; C7-3 D. Jarvis; C7-4 MRU student work; C7-5 D. Jarvis; C7-6 I Naested; C7-7 D. Jarvis; C7-8 D. Jarvis; C7-9 D. Jarvis; C7-10 D. Jarvis; C7-11 D. Jarvis; C7-12 D. Jarvis; C7-13 D. Jarvis; C7-14 Shutterstock: Dodecahedron Net, Image ID: 68245720 , Copyright: MilanB; C7-15 D. Jarvis; C7-16 D. Jarvis; C7-17 D. Jarvis; C7-18 D. Jarvis; C7-19 MRU student work; C7-20 D. Jarvis; C7-21 Shutterstock: Painted Eggs for the Orthodox Easter, Image ID: 26908447, Copyright: Andrei_Badau; C7-22 D. Jarvis; C7-23 D. Jarvis; C7-24 I. Naested; C7-25 D. Jarvis

CHAPTER 8: C8-1A student work, courtesy D. Jarvis, used with permission; C8-1B student work, courtesy D. Jarvis, used with permission; C8-1C student work, courtesy D. Jarvis, used with permission; C8-1D student work, courtesy D. Jarvis, used with permission; C8-1E student work, courtesy D. Jarvis, used with permission; C8-1F student work, courtesy D. Jarvis, used with permission

About the Authors

Dr. Daniel Jarvis is an Associate Professor and Chair of Graduate Studies in Education at the Schulich School of Education at Nipissing University, North Bay, Ontario, Canada. Having a background in both mathematics and visual arts, he has enjoyed teaching courses at all three levels of education. His research interests include integrated curricula, instructional technologies, teacher professional learning, and educational leadership.

Dr. Irene Naested is a Full Professor and Chair of the Department of Education and Schooling at Mount Royal University, Calgary, Alberta, Canada. Irene's background as an artist, author, and educational leader in K–12 and post-secondary institutions has provided her with many valuable, varied, and rich experiences. She is passionate with regard to the role of arts in the education of young learners.